T0137865

Studies in Systems, Decision and Control

Volume 218

The series "Studies in Systems, Decision and Control" (SSDC) covers both new developments and advances, as well as the state of the art, in the various areas of broadly perceived systems, decision making and control–quickly, up to date and with a high quality. The intent is to cover the theory, applications, and perspectives on the state of the art and future developments relevant to systems, decision making, control, complex processes and related areas, as embedded in the fields of engineering, computer science, physics, economics, social and life sciences, as well as the paradigms and methodologies behind them. The series contains monographs, textbooks, lecture notes and edited volumes in systems, decision making and control spanning the areas of Cyber-Physical Systems, Autonomous Systems, Sensor Networks, Control Systems, Energy Systems, Automotive Systems, Biological Systems, Vehicular Networking and Connected Vehicles, Aerospace Systems, Automation, Manufacturing, Smart Grids, Nonlinear Systems, Power Systems, Robotics, Social Systems, Economic Systems and other. Of particular value to both the contributors and the readership are the short publication timeframe and the world-wide distribution and exposure which enable both a wide and rapid dissemination of research output.

Indexed by SCOPUS, DBLP, WTI Frankfurt eG, zbMATH, SCImago.

All books published in the series are submitted for consideration in Web of Science.

Laxman Bokati · Vladik Kreinovich

Decision Making Under Uncertainty, with a Special Emphasis on Geosciences and Education

Laxman Bokati
Computational Science Program
The University of Texas at El Paso
El Paso, TX, USA

Vladik Kreinovich
Department of Computer Science
The University of Texas at El Paso
El Paso, TX, USA

ISSN 2198-4182 ISSN 2198-4190 (electronic)
Studies in Systems, Decision and Control
ISBN 978-3-031-26088-9 ISBN 978-3-031-26086-5 (eBook)
https://doi.org/10.1007/978-3-031-26086-5

This Springer imprint is published by the registered company Springer Nature Switzerland AG
The registered company address is: Gewerbestrasse 11, 6330 Cham, Switzerland

Contents

Part I
Introduction

Chapter 1
General Introduction

One of the main objective of science. One of the main objectives of science is to help people make good decisions. Because of the ubiquity and importance of decision making, it has been the subject of intensive research. This research can be roughly divided into two categories:

- analysis of how a rational person *should* make a decision, and
- analysis of how people *actually* make decisions.

The main objective of this dissertation is to expand on both research categories, with the ultimate objective of providing the corresponding practical recommendations.

Decision making under uncertainty. One of the major difficulties in decision making is that usually, we do not have full information about the situation and about possible consequences of our decisions. The larger this uncertainty, the more difficult it is to make decisions.

Selecting case studies. Since the larger the uncertainty, the more difficult it is to make a right decision, a natural case study for new decision making techniques are situations where decisions are the most difficult—i.e., where there is the largest amount of uncertainty. In most practical problems, even if we do not have the full information about the situation, i.e., even if we do not know the values of some quantities, we can, in principle, measure these values and get a better understanding of the situation. For example, while we often do not have enough information about the weather—i.e., about the current values of temperature, humidity, wind speeds etc. at different locations and different heights—we can, if needed, measure these values and thus, decrease the uncertainty.

There is, however, an application area where such measurements are not possible: namely, geosciences. For example, oil companies would like to know whether it makes sense to start digging an oil well at a prospective location. When we make this

L. Bokati and V. Kreinovich, *Decision Making Under Uncertainty, with a Special Emphasis on Geosciences and Education*, Studies in Systems, Decision and Control 218,
https://doi.org/10.1007/978-3-031-26086-5_1

decision, we do not have full information on what is happening at the corresponding depths, what is the density there, what is the speed of sound at this level, what are other physical characteristics. In principle, it is possible to perform direct measurements that can determine this information—but this measurement would require, in effect, digging a deep well and placing instruments down below, while the whole purpose of this analysis is to decide whether it is worth investing significant resources in this possible well in the first place. Because of this, geosciences are among the most challenging areas for decision making. Because of this, we have selected geosciences as the main case study for our results.

Another area where measurements are difficult is teaching. In education, we can gauge the observable results of teaching but not the internal process that lead to more or less successful teaching—just like in geosciences, we can measure the seismic waves reaching the surface, but we cannot directly measure the processes leading to these waves.

Structure of the dissertation. In line with all this, this dissertation is organized as follows.

Part 1. In the first—introductory—part, we provide a brief reminder of decision theory—a theory that explains how rational people should make decisions. The main ideas related to (rational) individual decision making are described in Chap. 2. Chapter 3 covers the general ideas behind (rational) group decision making. While the corresponding formulas are known, this chapter already contains some new material—namely, we provide a new simplified derivation of these formulas.

Finally, Chap. 4 explains how we can control group decision making by modifying the proposed options. This chapter contains both an empirical dependence—and our explanation of this dependence. This is the first chapter that contains completely new results.

We then follow with other parts that contain completely new material.

Part 2. First comes Part 2, in which:

- we analyze how people actually make decisions—in general and, in particular, in economy-related situations, and
- we explain why people's actual decisions differ from recommendations of decision theory.

This part covers all possible deviations of actual decisions from the ideal ones. In the ideal case:

1. first, we find the exact value of each item in each alternative,
2. then, we combine these values into exact values of each alternative,
3. after that, we find future consequences of different actions, and preferences of other people, and
4. finally, based on all this information, we select the optimal alternative.

In real life, human decision making deviates from the ideal on all these four stages.

1. On the first stage, we have to base our decisions on incomplete, approximate knowledge:

   - either because information leading to more accurate estimates are not available,
   - or because, while this information is available, there is not enough time to process this information.

   In such case:

   - Instead of coming up with the exact values of each item, people come up with approximate estimates—i.e., in effect, bounds on possible values. As we show in Chap. 5, this explains the empirical fact that people's selling prices are usually higher than their buying prices—the fact that seems to contradict the basic economic ideas.
   - Also, since the information is usually incomplete, different people come up with different prices for the same item—which explains the constant buying and selling, something that also seems to contradict the basic economic ideas; see Chap. 6.
   - Instead (or in lieu) of eliciting the accurate values, people make decisions based on clusters containing the actual values—e.g., use the so-called 7 ± 2 approach; see Chaps. 7 and 8.

2. On the second stage, when people combine utility values, they use approximate processing techniques; see, e.g., Chap. 9.
3. On the third stage, people use biased perceptions of the future time; see Chap. 10, resulting in non-optimal solutions (Chap. 11). They also have a biased perception of other people's utility, which also leads to non-optimal solutions; see Chap. 12.
4. Finally, on the fourth stage, instead of going for an optimal solution, people use approximately optimal solutions (Chap. 13) or even use heuristics instead of looking for optimal or approximately optimal solutions (Chaps. 14 and 15).

In most of these cases, there are known empirical formulas describing actual human behavior. In our analysis, we provide possible theoretical explanations for these formulas.

Part 3. After this general description of human decision making, in Part 3, we focus on our main application area: geosciences. In geosciences, like in many other application areas, we encounter two types of situations.

In some cases, we have a relatively small number of observations—only sufficiently many to estimate the values of a few parameters of the model. In such cases, it is desirable to come up with the most adequate few-parametric model. We analyze the corresponding problem of selecting an optimal model on two examples:

- of spatial dependence (Chap. 16) and
- of temporal dependence (Chap. 17).

As an example of a temporal dependence problem, we consider one of the most challenging and the most important geophysical problems: the problem of earthquake prediction. Specifically, we analyze the problem of selecting the most adequate probabilistic distribution of between-earthquakes time intervals.

In other cases, we already have many observations covering many locations and many moments of time. In such cases, we can look for the best ways to extend this knowledge:

- to other spatial locations (Chap. 18) and
- to future moments of time (Chap. 19).

As an example of extending knowledge to future moments of time—i.e., prediction—we deal with one of the least studied seismic phenomena: earthquakes triggering other earthquakes.

Part 4. In Part 4, we study applications to teaching. Our analysis cover all three related major questions:

- what to teach (Chaps. 20 and 21),
- how to teach (Chap. 22), and
- how to grade, i.e., how to gauge the results of teaching (Chap. 23).

Part 5. Most of these and other applications involve intensive computing. In the final Part 5, we show that the above-analyzed ideas can be used in all aspects of computing:

- in analyzing the simplest (linear) models (Chap. 24),
- in analyzing more realistic non-linear models (Chap. 25), and even
- in exploring perspective approaches to computing (Chap. 26).

Appendix. In all these parts, several of our applications are based on common (or at least similar) mathematical results. These results are summarized in a special mathematical Appendix.

Chapter 2
(Rational) Individual Decision Making: Main Ideas

What is traditional decision theory. Traditional decision theory (see, e.g., [1–5]) describes preferences of rational agents.

There are many aspects of rationality: e.g., if a rational agent prefers A to B and B to C, then this agent should prefer A to C.

Comment. As we have mentioned in the Introduction, preferences of real agents are not always rational in the above sense; see, e.g., [6, 7]. One of the main reasons for this deviation from rationality is that our ability to process information and select an optimal decision is bounded. However, in many cases, traditional decision theory still provides a very good description of human behavior.

The notion of utility. To describe the preferences of a rational agent, decision theory requires that we select two alternatives:

- a very bad one A_- that is much worse than anything this agent will actually encounter, and
- a very good one A_+ that is much better than anything this agent will actually encounter.

For each value p from the interval $[0, 1]$, we can then form a lottery $L(p)$ in which:

- we get the very good alternative A_+ with probability p and
- we get the very bad alternative A_- with the remaining probability $1 - p$.

When the probability p is close to 1, this means that we are almost certainly getting a very good deal. So, for any realistic option A, the corresponding lottery $L(p)$ is better than A: $A < L(p)$.

Similarly, when the probability p is close to 0, this means that we are almost certainly getting a very bad deal, so $L(p) < A$.

L. Bokati and V. Kreinovich, *Decision Making Under Uncertainty, with a Special Emphasis on Geosciences and Education*, Studies in Systems, Decision and Control 218,
https://doi.org/10.1007/978-3-031-26086-5_2

There should be a threshold u at which the preference $L(p) < A$ corresponding to smaller probabilities p is replaced by an opposite preference $A < L(p)$. In other words, we should have:

- $L(p) < A$ for all $p < u$ and
- $A < L(p)$ for all $p > u$.

This threshold value is called the *utility* of the alternative A. It is usually denoted by $u(A)$.

The above two conditions mean that, in a certain reasonable sense, the original alternative A is equivalent to the lottery $L(u(A))$ corresponding to the probability $u(A)$. We will denote this equivalence by $A \equiv L(u(A))$.

A rational agent should maximize utility. Of course, the larger the probability of getting a very good outcome A_+, the better. Thus, among several lotteries $L(p)$, we should select the one for which the probability p of getting the very good alternative A_+ is the largest.

Since each alternative A is equivalent to the corresponding lottery $L(u(A))$, this implies that we should select the alternative with the largest possible value of utility.

Main conclusion of the traditional decision theory: a rational agent must maximize expected utility. In practice, we rarely know the consequences of each action. At best, we know possible outcomes $A_1, \ldots, A_n$, and their probabilities $p_1, \ldots, p_n$.

Since each alternative A_i is equivalent to a lottery $L(u(A_i))$ in which we get A_+ with probability $u(A_i)$ and A_- with the remaining probability $1 - u(A_i)$, the whole action is therefore equivalent to the following two-stage lottery:

- first, we select one of the n alternatives A_i with probability p_i, and
- then, depending on which alternative A_i we selected on the first stage, we select A_+ with probability $u(A_i)$ and A_- with the remaining probability $1 - u(A_i)$.

As a result of this two-stage lottery, we get either the very good alternative A_+ or the very bad alternative A_-. The probability u of getting A_+ can be computed by using the formula of complete probability: it is equal to

$$u = p_1 \cdot u(A_1) + \cdots + p_n \cdot u(A_n).$$

One can see that this is exactly the formula for the expected value of the utility $u(A_i)$. Thus, the utility of each action to a person is equal to the expected value of utility.

So, according to the traditional decision theory, rational agents should select the alternative with the largest possible value of expected utility.

Utility is defined modulo linear transformations. The numerical value of utility depends on the selection of the alternatives A_- and A_+. It can be shown that if we select a different pair (A'_-, A'_+), then the corresponding utility $u'(A)$ is related to the original utility by a linear transformation $u'(A) = a \cdot u(A) + b$ for some $a > 0$ and b; see, e.g., [2, 4].

How utility is related to money. The dependence of utility of money is non-linear: namely, utility u is proportional to the square root of the amount m of money $u = c \cdot \sqrt{m}$; see [6] and references therein.

Comment. This empirical fact can be explained. For example, the non-linear character of this dependence is explained, on a commonsense level, in [8], while the square root formula can also be explained—but it requires more mathematical analysis; see, e.g., [7]. In the current book, we simply take this fact as a given.

How to compare current and future gains: discounting. How can we compare current and future gains? If we have an amount m of money now, then we can place it in a bank and get the same amount plus interest, i.e., get the new amount $m' \stackrel{\text{def}}{=} (1 + i) \cdot m$ in a year, where i is the interest rate. Thus, the amount m' in a year is equivalent to the value $m = q \cdot m'$ now, where $q \stackrel{\text{def}}{=} 1/(1 + i)$. This reduction of future gains—to make them comparable to current gains—is known as *discounting*.

Discounting: a more detailed description. An event—e.g., a good dinner—a year in the past does not feel as pleasant to a person now as it may have felt a year ago. Similarly, a not-so-pleasant event in the past—e.g., a painful inoculation—does not feel as bad now as it felt a year ago, when it actually happened. Thus, the utility of an event changes with time:

- positive utility decreases,
- negative utility increases,

and in both cases, the utility gets closer to its neutral value.

If u is the utility of a current event, how can we describe the utility $f(u)$ of remembering the same event that happened 1 year ago?

We can normalize the utility values by assuming that the status quo situation has utility 0. In this case, the starting point for measuring utility is fixed, and the only remaining transformation is re-scaling $u' = a \cdot u$.

It is reasonable to require that the function $f(u)$ is invariant with respect to such a transformation, i.e., that:

- if we have $v = f(u)$,
- then for each a, we should have $v' = f(u')$, where we denoted $v' = a \cdot v$ and $u' = a \cdot u$.

Substituting the expressions for v' and u' into the formula $v' = f(u')$, we conclude that $a \cdot v = f(a \cdot u)$, i.e., $a \cdot f(u) = f(a \cdot u)$. Substituting $u = 1$ into this formula, we conclude that $f(a) = q \cdot a$, where we denoted $q \stackrel{\text{def}}{=} f(1)$. Since $f(u) < u$ for $u > 0$, this would imply that $q < 1$.

- So, an event with then-utility u that occurred 1 year ago has the utility $q \cdot u$ now.

- Similarly, an event with utility u that happened 2 years ago is equivalent to $q \cdot u$ a year ago, and thus, is equivalent to $q \cdot (q \cdot u) = q^2 \cdot u$ now.

We can similarly conclude that an event with utility u that occurred t moments in the past is equivalent to utility $q^t \cdot u$ now.

Decision making under interval uncertainty. In real life, we rarely know the exact consequences of each action. As a result, for each alternative A, instead of the exact value of its utility, we often only know the bounds $\underline{u}(A)$ and $\overline{u}(A)$ on this unknown value. In other words, all we know is the interval $[\underline{u}(A), \overline{u}(A)]$ that contains the actual (unknown) value $u(A)$. How can we make a decision under this interval uncertainty?

In particular, for such an interval case, we need to be able to compare the interval-valued alternative with lotteries $L(p)$ for different values p. As a result of such comparison, we will come up with a utility of this interval. So, to make recommendations on decision under interval uncertainty, we need to be able to assign, to each interval $[\underline{u}, \overline{u}]$, a single utility value $u(\underline{u}, \overline{u})$ from this interval that describes this interval's utility.

Since utility is defined modulo a linear transformation $u \to u' = a \cdot u + b$, it is reasonable to require that the corresponding function $u(\underline{u}, \overline{u})$ should also be invariant under such transformations, i.e., that:

- if $u = u(\underline{u}, \overline{u})$,
- then $u' = u(\underline{u}', \overline{u}')$, where we denoted $u' = a \cdot u + b$, $\underline{u}' = a \cdot \underline{u} + b$, and $\overline{u}' = a \cdot \overline{u} + b$.

It turns out that this invariance requirement implies that

$$u(\underline{u}, \overline{u}) = \alpha_H \cdot \overline{u} + (1 - \alpha_H) \cdot \underline{u}$$

for some $\alpha_H \in [0, 1]$ [2, 4]. This formula was first proposed by a future Nobelist Leo Hurwicz and is, thus, known as the Hurwicz optimism-pessimism criterion [3, 9].

Theoretically, we can have values $\alpha_H = 0$ and $\alpha_H = 1$. However, in practice, such values do not happen:

- $\alpha_H = 1$ would correspond to a person who only takes into account the best possible outcome, completely ignoring the risk of possible worse situations;
- similarly, the value $\alpha_H = 0$ would correspond to a person who only takes into account the worst possible outcome, completely ignoring the possibility of better outcomes.

In real life, we thus always have $0 < \alpha_H < 1$.

References

1. P.C. Fishburn, *Utility Theory for Decision Making* (Wiley, New York, 1969)
2. V. Kreinovich, Decision making under interval uncertainty (and beyond), in *Human-Centric Decision-Making Models for Social Sciences*. ed. by P. Guo, W. Pedrycz (Springer, Berlin, 2014), pp.163–193
3. R.D. Luce, R. Raiffa, *Games and Decisions: Introduction and Critical Survey* (Dover, New York, 1989)
4. H.T. Nguyen, O. Kosheleva, V. Kreinovich, Decision making beyond Arrow's 'impossibility theorem', with the analysis of effects of collusion and mutual attraction. Int. J. Intell. Syst. **24**(1), 27–47 (2009)
5. H. Raiffa, *Decision Analysis* (McGraw-Hill, Columbus, Ohio, 1997)
6. D. Kahneman, *Thinking, Fast and Slow* (Farrar, Straus, and Giroux, New York, 2011)
7. J. Lorkowski, V. Kreinovich, *Bounded Rationality in Decision Making Under Uncertainty: Towards Optimal Granularity* (Springer, Cham, Switzerland, 2018)
8. O. Kosheleva, M. Afravi, V. Kreinovich, Why utility non-linearly depends on money: a commonsense explanation, in *Proceedings of the 4th International Conference on Mathematical and Computer Modeling* (Omsk, Russia, 11 Nov 2016), pp. 13–18
9. L. Hurwicz, *Optimality Criteria for Decision Making Under Ignorance*, Cowles Commission Discussion Paper, Statistics, No. 370 (1951)

References

1. [illegible]
2. [illegible]
3. [illegible]
4. [illegible]
5. [illegible]
6. [illegible]
7. [illegible]
8. [illegible]
9. [illegible]

Chapter 3
(Rational) Group Decision Making: General Formulas and a New Simplified Derivation of These Formulas

According to decision theory, if a group of people wants to select one of the alternatives in which all of them get a better deal than in a status quo situation, then they should select the alternative that maximizes the product of their utilities. This recommendation was derived by Nobelist John Nash. In this chapter, we describe this idea, and we also provide a new (simplified) derivation of this result, a derivation which is not only simpler—it also does not require that the preference relation between different alternatives be linear.

Comment. The result of this chapter first appeared in [1].

3.1 (Rational) Group Decision Making: General Formulas

Practical problem. In many practical situations, a group of people needs to make a joint decision. They can stay where they are—in the "status quo" state. However, they usually have several alternatives in which each of them gets a better deal than in the status quo state. Which of these alternatives should they select?

Nash's bargaining solution to this problem. To solve this problem, in 1950, Nobelist John Nash formulated several reasonable conditions that the selection must satisfy [2]. He showed that the only way to satisfy all these conditions is to select the alternative that maximizes the product of participants' *utilities*—special functions that describe a person's preferences in decision theory; see, e.g., Chap. 2 or [3–7].

What we do in this chapter. In this chapter, we provide a new simplified derivation of Nash's bargaining solution, a derivation that is not only simpler—this derivation also uses fewer assumptions: e.g., it does not assume that there is total (linear) preorder between different alternatives.

L. Bokati and V. Kreinovich, *Decision Making Under Uncertainty, with a Special Emphasis on Geosciences and Education*, Studies in Systems, Decision and Control 218,
https://doi.org/10.1007/978-3-031-26086-5_3

3.2 A New (Simplified) Explanation of Nash's Bargaining Solution

Let us describe the problem in precise terms. We consider a situation in which n participants need to make a joint decision. In this situation, each possible alternative can be characterized by a tuple $u = (u_1, \ldots, u_n)$ of the corresponding utility values. We only consider alternatives in which $u_i > 0$ for all i—otherwise why would the ith person agree to this alternative if he/she does not gain anything—or even lose something?

For some alternatives u and u', the group prefers u' to u. For example, if each person gets more in alternative u' than in u, then u' is clearly better than u. We will denote this preference by $u < u'$.

For some alternatives u and u', the group may consider them equally good; we will denote this by $u \sim u'$. We also allow the possibility that for some pairs of alternatives u and u', the group cannot decide which of them is better. In other words, we do not assume that the preference relation is total (linear).

The relations $<$ and $\sim$ should satisfy natural transitivity conditions: e.g., if u' is better than u and u'' is better than u', then u'' should be better than u. We will call such a pair of relations $(<, \sim)$ a *preference relation*; we will give a precise definition shortly.

We assume that the preference relation is *fair* in the sense that all participants are treated equally. In particular, if we perform any permutation of the utilities, the alternative should remain of the same quality to the group: e.g., (a, b) should be of the same quality as (b, a): $(a, b) \sim (b, a)$.

Finally, since utility u_i of each participants is determined modulo re-scaling $u_i \to c_i \cdot u_i$, relative preference of two different alternatives should not change if we perform such a re-scaling. Thus, we arrive at the following definitions.

Definition 3.1 Let A be a set; its elements will be called *alternatives*. By a *preference relation* on the set A, we mean a pair of relations $(<, \sim)$ with the following properties:

- for each a, we have $a \sim a$;
- for each a and b, if $a \sim b$, then $b \sim a$;
- for each a and b, if $a \sim b$, then we cannot have $a < b$;
- for each a, b, and c, if $a < b$ and $b < c$, then $a < c$;
- for each a, b, and c, if $a < b$ and $b \sim c$, then $a < c$;
- for each a, b, and c, if $a \sim b$ and $b < c$, then $a < c$;
- for each a, b, and c, if $a \sim b$ and $b \sim c$, then $a \sim c$.

Definition 3.2 Let $A = \mathbb{R}^n_+$ be a set of all n-tuples $u = (u_1, \ldots, u_n)$ on positive numbers, and let $(<, \sim)$ be a preference relation on the set A.

- We say that the pre-order is *monotonic* if whenever we have $u_i < u'_i$ for all i, then we should have $u < u'$.
- We say that the pre-order is *fair* if for each permutation

$$\pi : \{1, \ldots, n\} \to \{1, \ldots, n\}$$

and for each alternative u, we have $(u_1, \ldots, u_n) \sim (u_{\pi(1)}, \ldots, u_{\pi(n)})$.
- We say that the pre-order is *scale-invariant* if for every two alternatives u and u' and for each tuples $(c_1, \ldots, c_n)$ of positive numbers:
  - if $(u_1, \ldots, u_n) < (u'_1, \ldots, u'_n)$ then

$$(c_1 \cdot u_1, \ldots, c_n \cdot u_n) < (c_1 \cdot u'_1, \ldots, c_n \cdot u'_n);$$

  - if $(u_1, \ldots, u_n) \sim (u'_1, \ldots, u'_n)$ then

$$(c_1 \cdot u_1, \ldots, c_n \cdot u_n) \sim (c_1 \cdot u'_1, \ldots, c_n \cdot u'_n).$$

Proposition 3.1 *There is one and only one monotonic fair scale-invariant preference relation:*

$$(u_1, \ldots, u_n) < (u'_1, \ldots, u'_n) \Leftrightarrow \prod_{i=1}^{n} u_i < \prod_{i=1}^{n} u'_i;$$

$$(u_1, \ldots, u_n) \sim (u'_1, \ldots, u'_n) \Leftrightarrow \prod_{i=1}^{n} u_i = \prod_{i=1}^{n} u'_i.$$

Comments.

- So, we indeed have a new explanation for Nash's bargaining solution.
- Actually, the above preference relation has a stronger property of *strong monotonicity*: that if $u_i \le u'_i$ for all i and $u_i < u'_i$ for some i, then $u < u'$.

Proof It is easy to check that the preference relation corresponding to Nash's bargaining solution is indeed monotonic, fair, and scale-invariant. So, to complete the proof, it is sufficient to show that every monotonic, fair, and scale-invariant bargaining solution is indeed described by the above formulas.

Indeed, due to symmetry, for all values u_2, u_3, ..., u_n, we have

$$(1, u_2, u_3, \ldots, u_n) \sim (u_2, 1, u_3, \ldots, u_n).$$

For each value $u_1 > 0$, we can use scale-invariance with $c_1 = u_1$ and $c_2 = \cdots = c_n = 1$ and conclude that

$$(u_1, u_2, u_3, \ldots, u_n) \sim (u_1 \cdot u_2, 1, u_3, \ldots, u_n).$$

So, we replace two values u_1 and u_2 with the product $u_1 \cdot u_2$ and 1 without losing equivalence.

Similarly, we can replace the two values $u_1 \cdot u_2$ and u_3 with the product $(u_1 \cdot u_2) \cdot u_3$ and 1, so

$$(u_1 \cdot u_2, 1, u_3, u_4, \ldots, u_n) \sim (u_1 \cdot u_2 \cdot u_3, 1, 1, u_4, \ldots, u_n)$$

and thus, by transitivity—which is part of the definition of the preference relation—we get

$$(u_1, u_2, u_3, u_4, \ldots, u_n) \sim (u_1 \cdot u_2 \cdot u_3, 1, 1, u_4, \ldots, u_n).$$

We can then similarly absorb u_4, etc., until we get

$$(u_1, \ldots, u_n) = (u_1 \cdot \ldots \cdot u_n, 1, \ldots, 1).$$

So all alternatives with the same value of the product $u_1 \cdot \ldots \cdot u_n$ are equivalent to the same alternative

$$(u_1 \cdot \ldots \cdot u_n, 1 \ldots, 1)$$

and are, thus, equivalent to each other.

Because of this property, each alternative

$$(p, 1, \ldots, 1)$$

is equivalent to

$$\left(\sqrt[n]{p}, \sqrt[n]{p}, \ldots, \sqrt[n]{p}\right).$$

When $p < p'$, then $\sqrt[n]{p} < \sqrt[n]{p'}$. So, due to monotonicity,

$$\left(\sqrt[n]{p}, \sqrt[n]{p}, \ldots, \sqrt[n]{p}\right) < \left(\sqrt[n]{p'}, \sqrt[n]{p'}, \ldots, \sqrt[n]{p'}\right)$$

and thus,

$$(p, 1, \ldots, 1) < (p', 1, \ldots, 1).$$

So, indeed, alternatives with the larger value of the product are better.

The proposition is proven.

Comment. In the previous text, we dismissed the possibility of alternatives with some of the values 0. It turns out that this dismissal can also be justified on mathematical grounds. Namely, let us show that in this case, no preference relation can satisfy all the above requirements.

Proposition 3.2 *On the set $A = \mathbb{R}^n_{\ge 0}$ of all tuples with non-negative components, no preference relation is strongly monotonic, fair, and scale-invariant.*

Proof Due to symmetry, $(1, 0) \sim (0, 1)$. By using $c_1 = 2$ and $c_2 = 1$, we conclude that $(2, 0) \sim (0, 1)$, so by transitivity $(1, 0) \sim (2, 0)$, which contradicts to strong monotonicity. The proposition is proven.

3.3 Taking Empathy into Account

Utility in the traditional economic models. In the traditional economic models, it is usually assumed that a decision maker maximizes his/her gain (numerically expressed as utility u), and this utility value describes the effect of this decision on this person at this particular moment of time; see, e.g., [3–7].

Need to go beyond traditional models. In these models, person's decisions are not affected by gains (utilities) of others and/or by gains of the same person at future moments of time. To some extent this is true, but one can easily find examples where gains of others (and/or future gains of the same person) do affect our behavior.

Maybe a proverbial greedy capitalist would gladly earn an extra million by making his workers work more and thus, get less utility, but in general, hardly anyone would prefer, e.g., $101 to $100 if this increase is accompanied by someone's severe suffering. Some people spend all their money like there is no tomorrow and retire in poverty, but most people do limit somewhat their current expenses to save for retirement. It is all a matter of degree. Some people are not empathetic enough, some do not save enough—but to some degree, practically everyone is empathetic and practically everyone saves (at least something).

How to describe dependence on other's utilities. Let $u_i^{(0)}$ be approximate utilities that come only from this person's consumption. How can we describe the actual utilities u_i that take into account other people's feelings—i.e., in precise terms, other people's utilities?

A natural way is to add, to $u_i^{(0)}$, terms proportional to other people's utilities, i.e., to consider expressions of the type

$$u_i = u_i^{(0)} + \sum_{j \ne i} \alpha_{ij} \cdot u_j, \tag{3.1}$$

where each coefficient α_{ij} describes how the utility of the ith person depends on the utility of the jth person; see, e.g., [6, 8–16].

This phenomenon is known by a polite term *empathy*, since for positive values α_{ij}, this formula describes how people feel better if others around them are happier.

However, from the purely mathematical viewpoint, it is also possible to have negative values α_{ij}, when someone's happiness makes the other person unhappy. This is not just a mathematical example, such things like jealousy and hatred are, unfortunately, quite real :-(

References

1. H.P. Nguyen, L. Bokati, V. Kreinovich, New (simplified) derivation of Nash's bargaining solution. J. Adv. Comput. Intell. Intell. Inform. (JACIII) **24**(5), 589–592 (2020)
2. J. Nash, The bargaining problem. Econometrica **18**(2), 155–162 (1950)
3. P.C. Fishburn, *Utility Theory for Decision Making* (Wiley, New York, 1969)
4. V. Kreinovich, Decision making under interval uncertainty (and beyond), in *Human-Centric Decision-Making Models for Social Sciences*. ed. by P. Guo, W. Pedrycz (Springer, Berlin, 2014), pp.163–193
5. R.D. Luce, R. Raiffa, *Games and Decisions: Introduction and Critical Survey* (Dover, New York, 1989)
6. H.T. Nguyen, O. Kosheleva, V. Kreinovich, Decision making beyond Arrow's 'impossibility theorem', with the analysis of effects of collusion and mutual attraction. Int. J. Intell. Syst. **24**(1), 27–47 (2009)
7. H. Raiffa, *Decision Analysis* (McGraw-Hill, Columbus, Ohio, 1997)
8. G.S. Becker, *A Treatise on the Family* (Harvard University Press, Cambridge, Massachusetts, 1991)
9. T. Bergstrom, Love and spaghetti, the opportunity cost of virtue. J. Econ. Perspect. **3**, 165–173 (1989)
10. T. Bergstron, Systems of benevolent utility interdependence, University of Michigan, Technical report (1991)
11. B.D. Bernheim, O. Stark, Altruism within the family reconsidered: do nice guys finish last? Am. Econ. Rev. **78**(5), 1034–1045 (1988)
12. D.D. Friedman, *Price Theory* (South-Western Publ, Cincinnati, Ohio, 1986)
13. H. Hori, S. Kanaya, Utility functionals with nonpaternalistic intergerenational altruism. J. Econ. Theory **49**, 241–265 (1989)
14. A. Rapoport, Some game theoretic aspects of parasitism and symbiosis. Bull. Math. Biophys. **18** (1956)
15. A. Rapoport, *Strategy and Conscience* (New York, 1964)
16. F.J. Tipler, *The Physics of Immortality* (Doubleday, New York, 1994)

Chapter 4
How We Can Control Group Decision Making by Modifying the Proposed Options

In summary, this chapter explains how we can control group decision making by modifying the proposed options. This chapter contains both an empirical dependence—and our explanation of this dependence. This is the first chapter that contains completely new results.

Let us describe it in more detail. For each task, the larger the stimulus, the larger proportion of people agree to perform this task. In many economic situations, it is important to know how much stimulus we need to offer so that a sufficient proportion of the people will agree to perform the needed task. There is an empirical formula describing how this proportion increases as we increase the amount of stimulus. However, this empirical formula lacks a convincing theoretical explanation, as a result of which practitioners are somewhat reluctant to use it. In this chapter, we provide a theoretical explanation for this empirical formula, thus making it more reliable—and hence, more useable.

Comment. The results of this chapter first appeared in [1].

4.1 Formulation of the Problem

The larger the stimulus, the more people agree to do the task. In economics, we need to entice people to perform certain tasks—whether it is planting crops or working in a factory or writing a software package. When the corresponding stimulus is too small, no one will agree to perform the task. When the stimulus is very high, everyone will agree. The proportion p of people who agree to perform a task will increase with the increase in the stimulus s.

L. Bokati and V. Kreinovich, *Decision Making Under Uncertainty, with a Special Emphasis on Geosciences and Education*, Studies in Systems, Decision and Control 218,
https://doi.org/10.1007/978-3-031-26086-5_4

It is desirable to know the exact amount of stimulus. A company wants certain tasks to be performed, so it has to use some stimulus. It is therefore desirable to find the exact amount of stimulus needed:

- if the stimulus is too low, no one will volunteer,
- if it is very high, the tasks will be performed, but the company will lose too much money.

How the amount of stimulus is usually determined now. In many cases, the selection of the right stimulus is done mostly by trial and error. This is, e.g., how airline companies, in an overbooked situation, ask for volunteers to give up their seats and fly the next day: they increase the award offered to potential volunteers until they get enough volunteers.

Formulas are needed, and there are such formulas. Trial-and-error is a lengthy process, difficult to predict. It is therefore desirable to have some analytical expressions that would help us select the right amount of stimulus.

Such expressions exist; see, e.g., [2] and references therein. The most empirically adequate expression is

$$p = \frac{s^q}{s^q + c} \tag{4.1}$$

for some constants q and c.

This formula is purely empirical. One of the main limitation of this formula is that is it purely empirical, it does not have a convincing theoretical explanation. Practitioners are usually very suspicious of best-fit purely empirical formulas, they are reluctant so use these formulas, they prefer formulas for which some theoretical explanation exists—since purely empirical formulas often turn out to be wrong.

And in economics and related areas, such later-wrong empirical formulas are ubiquitous:

- when a country has a boom, empirical formulas predict exponential growth forever;
- when, in the 1920s, the number of telephone operators started growing exponentially, empirical formulas predicted that in a few decades, half of the population will be telephone operators;

there are many examples like that.

It is thus desirable to come up with a theoretical explanation for empirical formulas.

What we do in this chapter. In this chapter, we provide a theoretical explanation for the formula (4.1).

4.2 Main Idea and the Resulting Explanation

Let us reformulate the problem in terms of probabilities. In the above text, we talked about proportion of people who take on the task. From the mathematical viewpoint, a proportion is not something about which we know much.

But what is proportion? It is simply the probability that a randomly selected person will take on the task. So, whatever we said about proportions can be reformulated in terms of probabilities—and about probabilities, we know a lot!

Comment. Good news is that we do not even need to change the notation p, since both words "proportion" and "probability" start with the same letter p.

What do we know about probabilities? One of the most widely used facts about probabilities is that if we add new evidence E, the probability of each hypothesis H_i changes according to the Bayes formula (see, e.g., [3]), from the original value $p_0(H_i)$ to the new value

$$p(H_i \mid E) = \frac{p_0(H_i) \cdot p(E \mid H_i)}{\sum\limits_j p_0(H_j) \cdot p(E \mid H_j)}. \tag{4.2}$$

In our case, we have two hypotheses:

- the hypothesis H_0 that the person will take on the task, whose probability is $p(H_0)$, and
- the hypothesis H_1 that the person will not take on the task; its probability is equal to $p(H_1) = 1 - p(H_0)$.

In this case, the general formula (4.2) takes the form

$$p(H_0 \mid E) = \frac{p_0(H_0) \cdot p(E \mid H_0)}{p_0(H_0) \cdot p(E \mid H_0) + (1 - p_0(H_0)) \cdot p(E \mid H_1)} =$$

$$\frac{p(H_0) \cdot p(E \mid H_0)}{p(H_0) \cdot (p(E \mid H_0) - p(E \mid H_1)) + p(E \mid H_1)}.$$

In other words, the change of the probability from the previous value $p = p_0(H_0)$ to the new value $p' = p(H_0 \mid E)$ is described by the formula

$$p' = \frac{p \cdot p(E \mid H_0)}{p \cdot (p(E \mid H_0) - p(E \mid H_1)) + p(E \mid H_1)}. \tag{4.3}$$

If we divide both the numerator and the denominator of the formula (4.3) by $p(E \mid H_1)$, then we get the following expression:

$$p' = \frac{p \cdot \dfrac{p(E \mid H_0)}{p(E \mid H_1)}}{1 + p \cdot \left(\dfrac{p(E \mid H_0)}{p(E \mid H_1)} - 1\right)},$$

i.e., the expression

$$p' = \frac{a \cdot p}{1 + (1 - a) \cdot p}, \tag{4.4}$$

where we denoted

$$a \stackrel{\text{def}}{=} \frac{p(E \mid H_0)}{p(E \mid H_1)}.$$

In other words, from the mathematical viewpoint, the change of the probability from the previous value p to the new value p' is thus described by a fractional-linear formula (4.4).

Here comes our idea. Our idea is that when we increase the stimulus, the resulting change of the probability should follow the formula (4.4).

How can we formalize this idea. What does it mean "increase the stimulus"? Intuitively, it means that we increase all the previous stimuli the same way.

What does that mean? If we simply add \$10 to all the previous stimulus values, this does not mean that we increases all the stimuli the same way. For example:

- if the previous stimulus was \$5, this is a drastic 3-times increase, but
- if the previous stimulus was \$1000, this is a barely noticeable 1% increase.

From the economic viewpoint, it makes more sense to increase all the previous stimulus values proportionally; e.g.:

- increase all the values by 1%, or
- increase all the values by 10%, or
- increase all the values by a factor of three.

With such an increase, instead of previous stimulus value s, we get a new stimulus value $\lambda \cdot s$, where, e.g.:

- an over-the-board 1% increase means $\lambda = 1.01$,
- an over-the-board 10% increase means $\lambda = 1.1$, and
- an over-the-board 3-times increase means $\lambda = 3$.

In these terms, the main idea takes the following form.

Resulting formulation. We want to find an increasing function $p(s)$ for which $p(0) = 0$, $p(s) \to 1$ as $s \to \infty$, and for every $\lambda > 0$, there exists $a(\lambda)$ for which, for all s, we have

$$p(\lambda \cdot s) = \frac{a(\lambda) \cdot p(s)}{1 + (a(\lambda) - 1) \cdot p(s)}. \tag{4.5}$$

Our main result. Our main result is that every function $p(s)$ satisfying the above conditions has exactly the form (4.1), for some values q and c.

This is exactly what we wanted. Thus, we indeed have the desired justification of the empirical formula (4.1).

4.3 Proof of the Main Result

Let us reformulate the formula (4.5) **in terms of odds.** For this proof, it is convenient to reformulate probabilities p in terms of the *odds*, i.e., in terms of the ratio

$$o = \frac{p}{1-p}.$$

Let us first find the odds corresponding to the new probability $p(\lambda \cdot s)$. From the formula (4.5), we get

$$1 - p(\lambda \cdot s) = 1 - \frac{a(\lambda) \cdot p(s)}{1 + (a(\lambda) - 1) \cdot p(s)} =$$

$$\frac{1 + a(\lambda) \cdot p(s) - p(s) - a(\lambda) \cdot p(s)}{1 + (a(\lambda) - 1) \cdot p(s)} = \frac{1 - p(s)}{1 + (a(\lambda) - 1) \cdot p(s)}. \quad (4.6)$$

Dividing (4.5) by (4.6), we get

$$d(\lambda \cdot s) = \frac{p(\lambda \cdot s)}{1 - p(\lambda \cdot s)} = \frac{a(\lambda) \cdot p(s)}{1 - p(s)} = a(\lambda) \cdot \frac{p(s)}{1 - p(s)}.$$

Since the ratio in the right-hand side is exactly the odds $o(s)$ corresponding to the probability $p(s)$, we thus conclude that

$$o(\lambda \cdot s) = a(\lambda) \cdot o(s). \quad (4.7)$$

Now, we can use the known solution to the functional equation (4.7). According to [4], every monotonic solution of the equation (4.7) has the form

$$o(s) = C \cdot s^q \quad (4.8)$$

for some values C and q.

The general proof of this statement is somewhat complicated, but it becomes very straightforward if we make an additional natural assumption that the function $p(s)$ is differentiable. In this case, the ratio $o(s)$ is also differentiable. Due to the Eq. (4.7), the function $a(\lambda)$ is equal to the ratio of two differentiable functions

$$a(\lambda) = \frac{o(\lambda \cdot s)}{o(s)}$$

and is, thus, also differentiable. Thus, we can differentiate both sides of the Eq. (4.7) with respect to λ and get

$$s \cdot o'(\lambda \cdot s) = a'(\lambda) \cdot o(s),$$

where, as usual, $f'(x)$ denotes the derivative. In particular, for $\lambda = 1$, we get $s \cdot o'(s) = q \cdot o(s)$, where we denoted $q \stackrel{\text{def}}{=} a'(1)$. In other words, we have

$$s \cdot \frac{do}{ds} = q \cdot o.$$

We can separate the variables s and o if we divide both sides of the equation by $s \cdot o$ and multiply both sides by ds, then we get

$$\frac{do}{o} = q \cdot \frac{ds}{s}.$$

Integrating both sides, we get $\ln(o) = q \cdot \ln(s) + C_0$, where C_0 is an integration constant. By applying $\exp(x)$ to both sides, we then get $o(s) = C \cdot s^q$, where we denoted $C \stackrel{\text{def}}{=} \exp(C_0)$.

From the equality (4.8) **to the desired formula** (4.1). According to the formula (4.8), we have

$$o(s) = \frac{p(s)}{1 - p(s)} = C \cdot s^q.$$

By taking the inverse of both sides, we get

$$\frac{1 - p(s)}{p(s)} = 1 - \frac{1}{p(s)} = C^{-1} \cdot s^{-q},$$

thus

$$\frac{1}{p(s)} = 1 - C^{-1} \cdot s^{-q}$$

and therefore,

$$p(s) = \frac{1}{1 - C^{-1} \cdot s^{-q}}.$$

Multiplying both the numerator and the denominator by s^q, we get

$$p(s) = \frac{s^q}{s^q - C^{-1}}.$$

Probabilities are always smaller than or equal to 1, thus $s^q \le s^q - C^{-1}$, i.e., $C^{-1} < 0$. If we denote $c \stackrel{\text{def}}{=} -C^{-1}$, we will get the desired formula (4.1).

The main result is thus proven.

References

1. L. Bokati, V. Kreinovich, T.H Doan, How the proportion of people who agree to perform a task depends on the stimulus: a theoretical explanation of the empirical formula, in Nguyen Ngoc Thach, Doan Thanh Ha, Nguyen Duc Trung, V. Kreinovich (eds.), *Prediction and Causality in Econometrics and Related Topics*, (Springer, Cham, Switzerland, 2022), pp. 22–27
2. M. Khani, A. Ahmadi, H. Hajary, Distributed task allocation in multi-agent environments using cellular learning automata. Soft. Comput. **23**, 1199–1218 (2019)
3. D.J. Sheskin, *Handbook of Parametric and Non-Parametric Statistical Procedures* (Chapman & Hall/CRC, London, UK, 2011)
4. J. Aczel, J. Dhombres, *Functional Equations in Several Variables* (Cambridge University Press, Cambridge, UK, 1989)

Part II
How People Actually Make Decisions

In this part, we analyze how people actually make decisions—in general and, in particular, in economy-related situations, and we explain why people's actual decisions differ from recommendations of decision theory.

This part covers all possible deviations of actual decisions from the ideal ones. In the ideal case:

1. First, we find the exact value of each item in each alternative,
2. Then, we combine these values into exact values of each alternative,
3. After that, we find future consequences of different actions, and preferences of other people, and
4. Finally, based on all this information, we select the optimal alternative.

In real life, human decision making deviates from the ideal on all there three stages.

1. On the first stage, we have to base our decisions on incomplete, approximate knowledge:

   - either because information leading more accurate estimates are not available,
   - or because, while this information is available, there is not enough time to process this information.

   In such case:

   - Instead of coming up with the exact values of each item, people come up with approximate estimates—i.e., in effect, bounds on possible values. As we show in Chap. 5, this explains the empirical fact that people's selling prices are usually higher than their buying prices—the fact that seems to contradict the basic economic ideas.
   - Also, since the information is usually incomplete, different people come up with different prices for the same item—which explains the constant buying and selling, something that also seems to contradict the basic economic ideas; see Chap. 6.

- Instead (or in lieu of) eliciting the accurate values, people make decisions based on clusters containing the actual values–e.g., use the so-called 7 ± 2 approach; see Chaps. 7 and 8.

2. On the second stage, when people combine utility values, they use approximate processing techniques; see, e.g., Chap. 9.
3. On the third stage, people use biased perceptions of the future time; see Chap. 10, resulting in non-optimal solutions (Chap. 11). They also have a biased perception of other people's utility, which also leads to non-optimal solutions; see Chap. 12.
4. Finally, on the fourth stage, instead of going for an optimal solution, people use approximately optimal solutions (Chap. 13) or even use heuristics instead of looking for optimal or approximately optimal solutions (Chaps. 14 and 15).

In most of these cases, there are known empirical formulas describing actual human behavior. In our analysis, we provide possible theoretical explanations for these formulas.

Chapter 5
The Fact That We Can Only Have Approximate Estimates Explains Why Buying and Selling Prices are Different

5.1 People's Actual Decisions Often Differ from What Decision Theory Recommends

In the previous chapters, we explained how ideal agents should make decisions. Namely:

- first, we find the exact value of each item in each alternative,
- then, we combine these values into exact values of each alternative,
- after that, we find future consequences of different actions, and preferences of other people, and
- finally, based on all this information, we select the optimal alternative.

In many cases, real-life decisions deviate from these ideal recommendations. In this and following chapters, we describe these deviations, and we explain why people's actual decisions differ from the recommendations of decision theory. Let us start with the first stage of decision making. In this stage, we have to base our decisions on incomplete, approximate knowledge:

- either because information leading to more accurate estimates are not available,
- or because, while this information is available, there is not enough time to process this information.

In such case, instead of coming up with the exact values of each item, people come up with approximate estimates–i.e., in effect, bounds on possible values.

In this chapter, we explain that this approximate character of estimates explains the empirical fact that people's selling prices are usually higher than their buying prices–the fact that seems to contradict the basic economic ideas.

Comment. The results of this chapter first appeared in [1].

L. Bokati and V. Kreinovich, *Decision Making Under Uncertainty, with a Special Emphasis on Geosciences and Education*, Studies in Systems, Decision and Control 218,
https://doi.org/10.1007/978-3-031-26086-5_5

5.2 Buying and Selling Prices are Different: A Phenomenon and Its Current Quantitative Explanations

Buying and selling prices are different: a phenomenon. According to the naive understanding of economic behavior, we should decide, for ourselves, how much each object is worth to us. This worth amount should be the largest amount that we should be willing to pay if we are buying this object, and this same amount should be the smallest amount for which we should agree to sell this object.

However, in many experiments, the price participants are willing to pay to buy a certain item is different from the price they are willing to accept to part with this item. For example, students are willing to pay $3 for a mug but require to be paid at least $7 to sell it back. In other words, people estimate the consequences of losing an object differently than the consequences of gaining the same object; see, e.g., [2, 3] and references therein.

Current explanations of this phenomenon. The current explanation of this phenomenon is based on the fact that people are not clear on the value of each object. Instead of the exact monetary amount, at best, they have a range $[\underline{u}, \overline{u}]$ of possible values of this object's worth; see, e.g., [4, 5].

Need for a more detailed analysis. While [4, 5] provide a qualitative explanation for the loss aversion phenomenon, it is desirable to extend this to a quantitative analysis, an analysis that takes into account known results about rational decision making under interval uncertainty. This is what we do in this chapter.

5.3 A New (Hopefully, More Adequate) Quantitative Explanation

Decision making under interval uncertainty: case of monetary values. How can we make a decision if, instead of the exact value of an object, we only know the interval $[\underline{u}, \overline{u}]$ of possible values? In other words, what is the value $u(\underline{u}, \overline{u})$ that we are willing to pay for this object?

Clearly, since we know that the object is worth at least $\underline{u}$ and at most $\overline{u}$, this means that the price $u(\underline{u}, \overline{u})$ that we are willing to pay should also be at least $\underline{u}$ and at most $\overline{u}$:

$$\underline{u} \le u(\underline{u}, \overline{u}) \le \overline{u}. \tag{5.1}$$

This property is known as *boundedness*.

Another reasonable requirement is that if we have two different objects, with values in $[\underline{u}, \overline{u}]$ and $[\underline{v}, \overline{v}]$, then the price that we are willing to pay to buy both should be equal to the prices that we pay for each of them. Let us describe this second requirement in precise terms.

When we get two objects together, the smallest possible value of our purchase is when both objects are worth their smallest amounts $\underline{u}$ and $\underline{v}$. In this case, the overall worth of both objects is equal to the sum $\underline{u} + \underline{v}$. Similarly, the largest possible value of our purchase is when both objects are worth their largest amounts $\overline{u}$ and $\overline{v}$. In this case, the overall worth of both objects is equal to the sum $\overline{u} + \overline{v}$. Thus, for two objects sold together the interval of possible worth values is $[\underline{u} + \underline{v}, \overline{u} + \overline{v}]$. So, the second requirement takes the following form:

$$u(\underline{u} + \underline{v}, \overline{u} + \overline{v}) = u(\underline{u}, \overline{u}) + u(\underline{v}, \overline{v}). \tag{5.2}$$

This property is known as *additivity*.

It turns out (see, e.g., [6]) that the only functions that satisfy both requirements (5.1) and (5.2) are functions of the type

$$u(\underline{u}, \overline{u}) = \alpha_H \cdot \overline{u} + (1 - \alpha_H) \cdot \underline{u}, \tag{5.3}$$

for some $\alpha_H \in [0, 1]$. This fact easily follows from the fact that all bounded additive functions are linear; see, e.g., [7].

As we have mentioned in Chap. 2, the formula (5.3) was first proposed–for the case of utilities–by a future Nobelist Leo Hurwicz and is thus known as Hurwicz optimism-pessimism criterion [8, 9]. We have just shown that a similar formula can be used to estimate monetary value under interval uncertainty.

Hurwicz criterion explains the difference between buy and sell prices. When we buy an object whose worth is between $\underline{u}$ and $\overline{u}$, the best possible gain is $\overline{u}$ and the worst possible gain is $\underline{u}$. Thus, according to the Hurwicz criterion, we should be willing to pay the amount u_b (*b* for *buy*) which is equal to

$$u_b = \alpha_H \cdot \overline{u} + (1 - \alpha_H) \cdot \underline{u}. \tag{5.4}$$

On the other hand, if we already own this object and we sell it, then our loss is between $-\overline{u}$ and $-\underline{u}$. The most optimistic estimate for our resulting state is $-\underline{u}$ and the most pessimistic estimate is $-\overline{u}$. In this case, according to the Hurwicz criterion, this is equivalent to the value of

$$\alpha_H \cdot (-\underline{u}) + (1 - \alpha_H) \cdot (-\overline{u}). \tag{5.5}$$

Thus, to compensate for this loss, we need to get the amount u_s (*s* for *sell*) that, when added to the value (5.5), will result in 0, i.e., the value

$$u_s = \alpha_H \cdot \underline{u} + (1 - \alpha_H) \cdot \overline{u}. \tag{5.6}$$

We can see that, in general, the expressions for the buy u_b and sell u_s prices are different. Indeed, the only time when the prices are equal, i.e., when $u_b = u_s$, is when

$$\alpha_H \cdot \underline{u} + (1 - \alpha_H) \cdot \overline{u} = \alpha_H \cdot \overline{u} + (1 - \alpha_H) \cdot \underline{u}.$$

Moving all the terms to the left-hand side and adding resulting coefficients at $\overline{u}$ and $\underline{u}$, we conclude that

$$(2\alpha_H - 1) \cdot \underline{u} - (2\alpha_H - 1) \cdot \overline{u} = 0,$$

i.e., $(2\alpha_H - 1) \cdot (\underline{u} - \overline{u}) = 0$. Since we consider the case when we have uncertainty, i.e., when $\underline{u} \neq \overline{u}$, we thus conclude that $2\alpha_H - 1 = 0$, i.e., that $\alpha_H = 0.5$.

So, only people with $\alpha_H = 0.5$ buy and sell at exactly the same price. For everyone else–who is even slightly more optimistic or even slightly less optimistic than $\alpha_H = 0.5$–the buy and sell prices are different, and this is exactly what we observe.

References

1. L. Bokati, V. Kreinovich, Decision theory can explain why buying and selling prices are different. J. Uncertain Syst. **13**(3), 189–192 (2019)
2. D. Kahneman, *Thinking, Fast and Slow* (Farrar, Straus, and Giroux, New York, 2011)
3. R.H. Thaler, *Misbehaving: The Making of Behavioral Economy* (W.W. Norton & Co., New York, 2015)
4. D. Gal, "Selling behavioral economics", *New York Times*, October 10, 2018, International Edition, p. 10
5. D. Gal, D.R. Rucker, Loss aversion, intellectual inertia, and a call for a more contrarian science: a reply to Simonson & Kivetz and Higgins & Liberman. J. Consum. Psychol. **28**(3), 533–539 (2018)
6. V. Kreinovich, Decision making under interval uncertainty (and beyond), in *Human-Centric Decision-Making Models for Social Sciences*. ed. by P. Guo, W. Pedrycz (Springer Verlag, 2014), pp.163–193
7. J. Aczél, J. Dhombres, *Functional Equations in Several Variables*, Cambridge University Press, 2008
8. L. Hurwicz, *Optimality Criteria for Decision Making Under Ignorance*, Cowles Commission Discussion Paper, Statistics, No. 370, 1951
9. R.D. Luce, R. Raiffa, *Games and Decisions: Introduction and Critical Survey* (Dover, New York, 1989)

Chapter 6
The "No Trade Theorem" Paradox

Another consequence of the fact that people have only approximate estimates of economics-related quantities is that different people come up with different prices for the same item.

In this chapter, we show that this fact explains the so-called "no trade theorem" paradox, one of the main challenges in foundations of finance–that constant buying and selling of stocks seems to contradict the traditional decision theory. Indeed, if an expert trader wants to sell some stock, that means that this trader believes that this stock will go down; however, the very fact that another expert trader is willing to buy it means that this other expert believes that the stock will go up. The fact that equally good experts have different beliefs should dissuade the first expert from selling–and thus, trades should be very rare. However, in reality, trades are ubiquitous.

In this chapter, we show that a detailed application of decision theory solves this paradox and explains how a trade can be beneficial to both seller and buyer. This application also explains a known psychological fact–that depressed people are usually more risk-averse.

Comment. Results from this chapter first appeared in [1].

6.1 "No Trade Theorem" and Why It is a Paradox

"No trade theorem" paradox. When a bank or a hedge fund wants to buy a stock, this means that professionals running this financial institution believe that, in the future, this stock will increase in price. This makes perfect sense until we realize that for this institution to be able to buy this stock, some other institution needs to be willing to sell it at this price–which means that professionals running that other institution must believe that, in the future, this stock will decrease in price.

L. Bokati and V. Kreinovich, *Decision Making Under Uncertainty, with a Special Emphasis on Geosciences and Education*, Studies in Systems, Decision and Control 218,
https://doi.org/10.1007/978-3-031-26086-5_6

Stock market is not a game for amateurs, serious agents buying and selling stock are smart experts who know what they are doing and who, in the past, have shown a good intuition about future stock values. So, even when such an expert initially thinks that this stock will increase in price, the very fact that this stock is available for sale means that another expert has an exactly opposite belief. This should, in many cases, dissuade the first expert from his or her original belief.

Similarly, an expert who is initially eager to sell, i.e., who initially believes that this stock will decrease in price, should be dissuaded by the presence of similarly qualified experts who are willing to buy, i.e., who believe that this stock will increase in price.

If we follow this logic, then very few agents will be trading stocks–but in reality, the trading volume is very high, every second, a huge amount of stocks change hands. This paradoxical behavior is known as a "no trade theorem"; see, e.g., [2, 3].

What we do in this chapter. In this chapter, we use decision making to show that in reality, trading makes perfect sense if we take into account different risks associated with different stocks.

Our explanation also explains another empirical phenomenon, a phenomenon from psychology–that depressed people are more risk-averse.

6.2 Analysis of the Problem and the Resulting Explanation of the "No Trade Theorem" Paradox

Towards formulation of the problem in precise terms. Let us assume that the person originally had the amount M of money. This person is thinking of possibly buying a stock which costs s.

Let us also assume an ideal situation, in which everyone has the same information about the future value of this stock, namely, everyone knows the probability distribution of its next year's gain. In particular, everyone knows the mean m' and the standard deviation σ' of this future gain. After discounting, we get the mean $m = q \cdot m'$ and the standard deviation $\sigma = q \cdot \sigma'$ of the equivalent current gain.

Let us also make a realistic assumption that the price s, the mean m, and the standard deviation σ are much smaller than the current money amount M. In other words, we assume that we are talking about a usual trade, not about extreme situations in which a person gambles his or her whole fortune by investing it all in a seemingly attractive stock.

Analysis of the problem. Let us denote the difference between the actual (discounted) value v of the stock and its mean value m by $\Delta v \stackrel{\text{def}}{=} v - m$. By definition of the mean, we have $m = E[v]$, where $E[\cdot]$ denoted the mean value. Thus, the mean value of Δv is 0: $E[\Delta v] = 0$.

The mean value of $(\Delta v)^2$ is, by definition, equal to σ^2: $E\left[(\Delta v)^2\right] = \sigma^2$.

In these terms, the (discounted) future gain is equal to $v = m + \Delta v$. The discounted future amount of money can be obtained if we take the original amount M, subtract the cost s of the stock, and add the gained value $v = m + \Delta v$; as a result, we get the value $M - s + m + \Delta v$.

The utility is proportional to the square root of money. We can always select a unit of utility so that utility will be exactly equal to the square root of money. In this case, the original utility is $u = \sqrt{M}$, and the discounted future utility corresponding to buying a stock is equal to $u = \sqrt{M - s + m + \Delta v}$.

As we have mentioned in the previous text, a rational person should select the alternative with the largest possible value of expected utility $E[u]$. Thus, for the agent, it makes sense to buy the stock if

$$E[u] = E\left[\sqrt{M - s + m + \Delta v}\right] > \sqrt{M}.$$

If we have a reverse inequality, then, as one can easily see, it is beneficial for this person to sell this stock. So, to decide whether it is beneficial for a person to sell or buy the stock, we need to estimate the value $E[u] = E\left[\sqrt{M - s + m + \Delta v}\right]$ of the expected utility.

Estimating the value of the expected utility. We assumed that the values m, s, and σ are much smaller than M. Thus, the corresponding random value Δv is also much smaller than M. So, we can expand the expression $\sqrt{M - s + m + \Delta v}$ in Taylor series in terms of s, m, and Δv, and keep only linear and quadratic terms in this expansion. As a result, we get the following expression:

$$\sqrt{M - s + m + \Delta v} = \sqrt{M} + \frac{1}{2\sqrt{M}} \cdot (-s + m + \Delta v) - \frac{1}{4 \cdot M^{3/2}} \cdot (-s + m + \Delta v)^2.$$

If we open the parentheses and take into account that $E[\Delta v] = 0$ and $E\left[(\Delta v)^2\right] = \sigma^2$, we conclude that the expected utility of buying the stock is equal to

$$E[u] = \sqrt{M} + \frac{1}{2\sqrt{M}} \cdot (m - s) - \frac{1}{4 \cdot M^{3/2}} \cdot ((m - s)^2 + \sigma^2).$$

Thus, this value is larger than the original utility $\sqrt{M}$ if and only if $E[u] - \sqrt{M} > 0$, i.e., if and only if

$$\frac{1}{2\sqrt{M}} \cdot (m - s) - \frac{1}{4 \cdot M^{3/2}} \cdot ((m - s)^2 + \sigma^2) > 0.$$

Multiplying both sides by $4 \cdot M^{3/2}$, we get an equivalent inequality

$$2M \cdot (m - s) - ((m - s)^2 + \sigma^2) > 0,$$

i.e., equivalently, $2M \cdot (m - s) > ((m - s)^2 + \sigma^2)$ and

$$M > M_0 \stackrel{\text{def}}{=} \frac{(m - s)^2 + \sigma^2}{2(m - s)}. \tag{6.1}$$

This explains the "no trade theorem" paradox. For the same stock with the same information about its future gains, whether it is beneficial to buy it or sell it depends on the initial amount of money that a trader has:

- if the trader has a large amount of money M, then buying a stock whose expected benefits m exceed the buying cost s makes perfect sense, even when the risk σ is reasonably high;
- on the other hand, if the trader has a not so large amount of money and the stock is risky, then for this trader, it makes sense to sell this stock.

For this stock, for almost all traders (with a rare exception of a trader whose current amount is exactly M_0), it is either beneficial to buy (if $M > M_0$) or to sell (if $M < M_0$).

Thus, for the same stock, with the same information, we always have many traders for whom it is beneficial to buy, and we have many traders for whom it is beneficial to sell. This explains the ubiquity of trading.

6.3 Auxiliary Result: Decision Theory Explains Why Depressed People are More Risk-Averse

Empirical fact. It has been observed that depressed people are more risk-averse, i.e., they are less willing to make decisions involving risks; see, e.g., [4, 5].

Our explanation. Each risky decision is described by the same formulas as a particular case of buying-a-stock risky decision: we may gain something, we may lose something, all we know is the probability distribution of the corresponding gains and losses.

Thus, to decide when it is beneficial to participate in a risky activity, we can use the same formulas as above–the only difference is that instead of just money amount M and the corresponding initial utility $u_0 = \sqrt{M}$, we can take into account different things that affect the person's utility. In terms of utility u_0, the inequality (6.1)–that describes when it is beneficial for a person to engage in a risky behavior–takes the form

$$u_0^2 > M_0 \stackrel{\text{def}}{=} \frac{(m - s)^2 + \sigma^2}{2(m - s)}. \tag{6.2}$$

This formula says that when the initial value of the utility u_0 is small, risky behavior–with large σ–is not beneficial. And this is exactly what depression means in decision-theoretic terms: that a person is not very happy, i.e., that the corresponding utility value u_0 is small.

Thus, our decision-theoretic analysis explains the above-mentioned psychological phenomenon.

References

1. L. Bokati, V. Kreinovich, Is 'No Trade Theorem' really a paradox: analysis based on decision theory. Appl. Math. Sci. **13**(8), 397–404 (2019)
2. P. Milgrom, N. Stokey, Information, trade and common knowledge. J. Econ. Theory. **26**(1), 17–27 (1982)
3. R.H. Thaler, *Misbehaving: The Making of Behavioral Economy* (W.W. Norton & Co., New York, 2015)
4. Y. Leykin, C.S. Roberts, R.J. DeRubeis, Decision-making and depressive symptomatology. Cogn. Ther. Res. **35**, 333–341 (2011)
5. F.C. Murphy, J.S. Rubinsztein, A. Michael, R.D. Rogers, T.W. Robbins, E.S. Paykel, B.J. Sahakian, Decision-making cognition in mania and depression. Psychol. Med. **31**, 679–693 (2001)

Chapter 7
People Make Decisions Based on Clusters Containing Actual Values

Instead of eliciting the accurate values, people make decisions based on clusters containing the actual values. It is known that, in general, people classify objects into 5 to 9 clusters—this is known as the 7 ± 2 law. In this chapter, we provide a possible simple geometric explanation for this psychological feature.

Comment. The results of this chapter first appeared in [1].

7.1 Formulation of the Problem

Phenomenon. There is a known phenomenon in psychology called a 7 ± 2 law (see, e.g., [2, 3]), according to which each person can directly keep in mind only a certain number of classes; depending on the person, this number ranges from $7 - 2 = 5$ to $7 + 2 = 9$ elements.

Why? A natural question is: why between 5 and 9? There have been some attempts to explain this phenomenon (see, e.g., [4]), but they are rather complex and not very intuitive.

In this chapter, we provide a possible geometric explanation for this phenomenon.

L. Bokati and V. Kreinovich, *Decision Making Under Uncertainty, with a Special Emphasis on Geosciences and Education*, Studies in Systems, Decision and Control 218,
https://doi.org/10.1007/978-3-031-26086-5_7

7.2 A Possible Geometric Explanation

Main idea. The above phenomenon is about our biological nature, so it has to be explained based on how it helped our ancestors survive. In order to survive in situations when there are dangerous and skilled predators around, it is important, for each person, to be aware of what is happening in the nearest vicinity.

Let us show that this natural idea indeed seems to explain the 7 ± 2 phenomenon.

Grid model. For simplicity, let us consider a simplified "grid" model of the environment, when the whole area is divided into square-shaped cells. In this model, instead of listing the exact spatial location of each object, we only describe in which cell this object is.

From this viewpoint, the space looks like this, with a person in the central cell marked by an X:

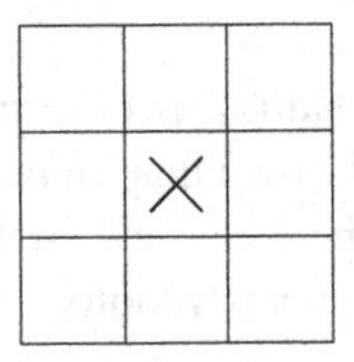

Awareness of nearest neighbors. For each person, it is vitally important to be aware of what is happening in the neighboring cells—so as not to miss a tiger or another dangerous predator nearby. From this viewpoint, it is important for a person standing in the middle of the above configuration to be aware of what is happening not only in the cell containing the person, but also in all the directly neighboring cells:

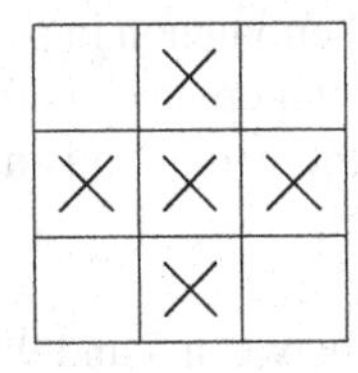

This requires keeping track of exactly five cells.

A better strategy. An even better strategy is to take into account not only directly neighboring cells, but also cells which are attached to the cell-where-we-are even by a single point—i.e., to take into account even the diagonally connected cells.

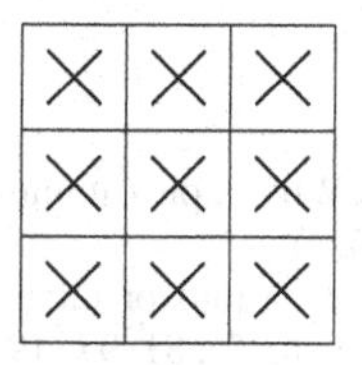

This requires keeping track of exactly nine cells.

Conclusion. To survive in a harsh environment, our ancestors had to keep track of the contents of between five and nine spatial cells. And this is exactly what we observe in the 7 ± 2 law—that we can keep track of between $7 - 2 = 5$ and $7 + 2 = 9$ objects.

7.3 Auxiliary Observation: How all This is Related to Our Understanding of Directions

How do we describe directions? In principle, we could divide the 360 degrees into 3, 4, 5, 6 parts.

How we actually navigate is that we use four main directions: South (S), North (N), East (E), and West (W).

This usual description of directions is related to the 5-neighboring-cells image. Together with the option to stay in the same place and not to move anywhere, we get the same 5-component picture as above:

N
W E
S

A more detailed description of directions is related to the 9-neighboring-cells image. A more detailed description of directions involves considering intermediate directions: Southwest (SW), Northwest (NW), Southeast (SE), and Northeast (NE). Together, we get the same 9-component picture as above:

NW N NE
W E
SW S SE

References

1. L. Bokati, V. Kreinovich, J. Katz, Why 7 plus minus 2? a possible geometric explanation. Geombinatorics **30**(1), 109–112 (2021)
2. G.A. Miller, The magical number seven plus or minus two: some limits on our capacity for processing information. Psychol. Rev. **63**(2), 81–97 (1956)
3. S.K. Reed, *Cognition: Theories and Application* (Wadsworth Cengage Learning, Belmont, California, 2010)
4. R. Trejo, V. Kreinovich, I.R. Goodman, J. Martinez, R. Gonzalez, A realistic (non-associative) logic and a possible explanations of 7 ± 2 law. Int. J. Approximate Reasoning **29**, 235–266 (2002)

Chapter 8
When Revolutions Succeed

A statistical analysis of hundreds of successful and unsuccessful revolution attempts led historians to a very unexpected conclusion: that most attempts involving at least 3.5% of the population succeeded, while most attempts that involved a smaller portion of the population failed. In this chapter, we show that this unexpected threshold can be explained based on the other two known rules of human behavior: the 80/20 rule (20% of the people drink 80% of the beer) and 7 plus minus 2 law—described in the previous chapter—according to which we naturally divide everything into 7 plus minus 2 classes.

Comment. Results from this chapter first appeared in [1].

8.1 Formulation of the Problem

Interesting empirical fact. Sometimes revolutions succeed, sometimes they don't. Researchers studying successful and unsuccessful revolutions usually go deep into each individual case, providing specific social, economic, and other explanations for the past success or failure.

A few years ago, researchers decided to analyze *all* successful and unsuccessful revolutions as a whole. The results of this analysis—presented in [2]—are somewhat unexpected: it turns out that there is a "magic" number of 3.5%:

- in the overwhelming majority of cases in which at least 3.5% of the population supported the revolution, this revolution won;
- on the other hand, in the overwhelming majority of cases in which less than 3.5% of the population supported the revolution attempt, this attempt failed.

L. Bokati and V. Kreinovich, *Decision Making Under Uncertainty, with a Special Emphasis on Geosciences and Education*, Studies in Systems, Decision and Control 218,
https://doi.org/10.1007/978-3-031-26086-5_8

This conclusion was very unexpected, since, contrary to what historians expected, the resulting "3.5% rule" does not depend on a social or economic situation, does not depend on the severity of the corresponding crisis, does not depend on the strategies of the revolutionaries and of the defenders of the old regime.

How can we explain this empirical law?

What we do in this chapter. In this chapter, we provide an explanation for this law—namely, we show that this law can be explained based on the other two well-known laws of human behavior: the 80/20 rule and the 7 plus minus 2 law. In the following text, we briefly recall these laws, and we explain how these laws imply the 3.5% rule.

8.2 80/20 Rule: Reminder

In most human activities, 80% of the results come from 20% of the participants; see, e.g., [3, 4]. For example:

- 20% of the people own 80% of all the world's wealth;
- 20% of researchers publish 80% of all the papers;
- 20% of the people contribute 80% of all the charitable donations; etc.,

not to mention the most frequently cited fact that 20% of the people drink 80% of the beer :-)

The usual explanation for this rule (see, e.g., [4] and references therein) is that the distribution of each of the corresponding quantities—money, papers, beer, etc.—follows the *power law*. To be more precise, if we sort people in decreasing order by the corresponding quantity, then the portion q of this quantity owned by the part p of the population is equal to $q = C \cdot p^\alpha$ for some C and α. Since for $p = 1$—when we consider the whole population—this portion should be 1, we conclude that $C = 1$ and thus,

$$q = p^\alpha. \tag{8.1}$$

8.3 How These Two Laws Explain the 3.5% Rule

The 80/20 rule describes a distribution of all possible quantities. In particular, it can be applied to the distribution of influence. We can therefore conclude that 20% of the people exert 80% of all the influence.

What about smaller portions than 20%? From the power law (8.1), we can make simple conclusions about the resulting influence. For example, if we square both sides of the formula (8.1), we conclude that $q' = (p')^\alpha$, where $q' = q^2$ and $p' = p^2$. In particular, if we apply this conclusion to our original example of

- $p = 20\% = 0.2$ and
- $q = 80\% = 0.8$,

we conclude that the same property holds for

- $p' = p^2 = 0.04 = 4\%$ and
- $q' = q^2 = 0.64 = 64\%$.

In other words, we come to a conclusion that if 4% of most active people act together, they exhibit 64% of all possible influence.

If we get a slightly smaller group of people—e.g., a group closer to 3.5% in size—they will still show about 60% of the overall influence.

What is so special about 60%? Why not 50%—a majority? The reason is very simple: if we have slightly more than 50%, we may not notice that we have a majority—that is why in close elections, when one side gets slightly more than 50% of the votes, there is often a bitter dispute in which both sides honestly and strongly believe that they have won. To be convincing, the majority must be convincing to even the least discriminatory people—i.e., the ones who divide everything into only 5 classes. For such persons to recognize the majority, this majority needs to constitute the majority of his/her 5 classes—i.e., at least 3 classes out of 5. This 3 out of 5 is exactly 60%—so the fact that $\approx$ 3.5% of the most active people have 60% of all the influence explains why this proportion of people is necessary for the revolution to succeed.

In short, a revolution succeeds if the active people involved in it exert the overwhelming majority of influence—overwhelming in the sense that even the least discriminatory people, who divided everything into only 5 classes, realize that yes, this is indeed a majority opinion. Thus, the 3.5% is indeed explained.

References

1. L. Bokati, O. Kosheleva, V. Kreinovich, When revolutions succeed? 80/20 rule and 7 plus minus 2 law explain the 3.5% rule. J. Uncertain Syst. **13**(3), 186–188 (2019)
2. E. Chenoweth, M.J. Stephan, *Why Civil Resistance Works: The Strategic Logic of Nonviolent Conflict* (Columbia University Press, New York, 2012)
3. P. Gomperts, A. Kovner, J. Lerner, D. Scharfstein, *Skill vs. Luck in Entrepreneurship and Venture Capital: Evidence from Serial Entrepreneurs*, US National Bureau of Economic Research (NBER) (2006) Working Paper 12592. Available online at: http://www.nber.org/papers/w12592
4. R. Koch, *The 80/20 Principle and 92 Other Powerful Laws of Nature: The Science of Success* (Nicholas Brealey, London, UK, 2014)

Chapter 9
How People Combine Utility Values

To make a decision, people need to estimate the value (utility) of each alternative. In real life, alternatives are complex, they include several different items. So, to estimate the value of an alternative, we need to estimate the value of each item, and then combine these estimates into a single number characterizing the alternative as a whole. In the previous chapters, we described how people estimate the value of each item. In this chapter, we will analyze how they combine these values into a single alternative-wide value.

In the ideal case, they should, e.g., add the monetary value of each item. In practice, however, their combination practices are different. These practices can be explained from the viewpoint of common sense. For example, if we place a can of coke that weighs 0.35 kg into a car that weighs 1 ton = 1000 kg, what will be the resulting weight of the car? Mathematics says 1000.35 kg, but common sense says 1 ton. In this chapter, we show that this common sense answer can be explained by the Hurwicz optimism-pessimism criterion of decision making under interval uncertainty.

Comment. This result was previously announced in [1].

9.1 Common Sense Addition

Suppose that we have two factors that affect the accuracy of a measuring instrument. One factor leads to errors $\pm 10\%$—meaning that the resulting error component can take any value from -10% to $+10\%$. The second factor leads to errors of $\pm 0.1\%$. What is the overall error?

From the purely mathematical viewpoint, the largest possible error is 10.1%. However, from the common sense viewpoint, an engineer would say: 10%.

L. Bokati and V. Kreinovich, *Decision Making Under Uncertainty, with a Special Emphasis on Geosciences and Education*, Studies in Systems, Decision and Control 218,
https://doi.org/10.1007/978-3-031-26086-5_9

As we have shown in the beginning of this chapter, similar common sense addition occurs in other situations as well. How can we explain this common sense addition?

9.2 Towards Precise Formulation of the Problem

We know that the overall measurement error Δx is equal to $\Delta x_1 + \Delta x_2$, where:

- the value Δx_1 can take all possible values from the interval $[-\Delta_1, \Delta_1]$, and
- the value Δx_2 can take all possible values from the interval $[-\Delta_2, \Delta_2]$.

What can we say about the largest possible value Δ of the absolute value $|\Delta|$ of the sum

$$\Delta x = \Delta x_1 + \Delta x_2?$$

Let us describe this problem in precise terms. For every pair (x_1, x_2):

- let $\pi_1(x_1, x_2)$ denote x_1 and
- let $\pi_2(x_1, x_2)$ stand for x_2.

Let $\Delta_1 > 0$ and $\Delta_2 > 0$ be two numbers. Without losing generality, we can assume that

$$\Delta_1 \geq \Delta_2.$$

By $\mathscr{S}$, let us denote the class of all possible sets

$$S \subseteq [-\Delta_1, \Delta_1] \times [-\Delta_2, \Delta_2]$$

for which

$$\pi_1(S) = [-\Delta_1, \Delta_1] \text{ and } \pi_2(S) = [-\Delta_2, \Delta_2].$$

We are interested in the value

$$\Delta(S) = \max\{|\Delta x_1 + \Delta x_2| : (\Delta x_1, \Delta_2) \in S\}$$

corresponding to the actual (unknown) set S.

We do not know what is the actual set S, we only know that $S \in \mathscr{S}$. For different sets $S \in \mathscr{S}$, we may get different values $\Delta(S)$. The only thing we know about $\Delta(S)$ is that it belongs to the interval $[\underline{\Delta}, \overline{\Delta}]$ formed by the smallest and the largest possible values of $\Delta(S)$ when $S \in \mathscr{S}$:

$$\underline{\Delta} = \min_{S \in \mathscr{S}} \Delta(S), \quad \overline{\Delta} = \max_{S \in \mathscr{S}} \Delta(S).$$

Which value Δ from this interval should we choose?

9.3 Hurwicz Optimism-Pessimism Criterion: Reminder

Situations when we do not know the value of a quantity, we only know the interval of its possible values, are ubiquitous. In such situations, decision theory recommends using *Hurwicz optimism-pessimism criterion*: selecting the value

$$\alpha \cdot \underline{\Delta} + (1 - \alpha) \cdot \overline{\Delta}$$

for some $\alpha \in [0, 1]$. A usual recommendation is to use $\alpha = 0.5$; see, e.g., [2–4].

Let us see what will be the result of applying this criterion to our problem.

9.4 Analysis of the Problem and the Resulting Explanation of Common Sense Addition

Computing $\overline{\Delta}$. For every set $S \in \mathscr{S}$, from $|\Delta x_1| \le \Delta_1$ and $|\Delta x_2| \le \Delta_2$, we conclude that

$$|\Delta x_1 + \Delta x_1| \le \Delta_1 + \Delta_2.$$

Thus always

$$\Delta(S) \le \Delta_1 + \Delta_2$$

and hence,

$$\overline{\Delta} = \max \Delta(S) \le \Delta_1 + \Delta_2.$$

On the other hand, for the set

$$S_0 = \{(v, (\Delta_2/\Delta_1) \cdot v) : v \in [-\Delta_1, \Delta_1]\} \in \mathscr{S},$$

we have

$$\Delta x_1 + \Delta x_2 = \Delta x_1 \cdot (1 + \Delta_2/\Delta_1).$$

Thus in this case, the largest possible value $\Delta(S_0)$ of $\Delta x_1 + \Delta x_2$ is equal to

$$\Delta(S_0) = \Delta_1 \cdot (1 + \Delta_2/\Delta_1) = \Delta_1 + \Delta_2.$$

So,

$$\overline{\Delta} = \max \Delta(S) \ge \Delta(S_0) = \Delta_1 + \Delta_2.$$

Hence,

$$\overline{\Delta} = \Delta_1 + \Delta_2.$$

Computing $\underline{\Delta}$. For every $S \in \mathscr{S}$, since

$$\pi_1(S) = [-\Delta_1, \Delta_1],$$

we have

$$\Delta_1 \in \pi_1(S).$$

Thus, there exists a pair

$$(\Delta_1, \Delta x_2) \in S$$

corresponding to

$$\Delta x_1 = \Delta_1.$$

For this pair, we have

$$|\Delta x_1 + \Delta x_2| \geq |\Delta x_1| - |\Delta x_2| = \Delta_1 - |\Delta x_2|.$$

Here, $|\Delta x_2| \leq \Delta_2$, so

$$|\Delta x_1 + \Delta x_2| \geq \Delta_1 - \Delta_2.$$

Thus, for each set $S \in \mathscr{S}$, the largest possible value $\Delta(S)$ of the expression

$$|\Delta x_1 + \Delta x_2|$$

cannot be smaller than $\Delta_1 - \Delta_2$:

$$\Delta(S) \geq \Delta_1 - \Delta_2.$$

Hence,

$$\underline{\Delta} = \min_{S \in \mathscr{S}} \Delta(S) \geq \Delta_1 - \Delta_2.$$

On the other hand, for the set

$$S_0 = \{(v, -(\Delta_2/\Delta_1) \cdot v) : v \in [-\Delta_1, \Delta_1]\} \in \mathscr{S},$$

we have

$$\Delta x_1 + \Delta x_2 = \Delta x_1 \cdot (1 - \Delta_2/\Delta_1).$$

Thus in this case, the largest possible value $\Delta(S_0)$ of $\Delta x_1 + \Delta x_2$ is equal to

$$\Delta(S_0) = \Delta_1 \cdot (1 - \Delta_2/\Delta_1) = \Delta_1 - \Delta_2.$$

So,

$$\underline{\Delta} = \min_{S \in \mathscr{S}} \Delta(S) \geq \Delta(S_0) = \Delta_1 - \Delta_2.$$

Thus,

$$\underline{\Delta} \le \Delta_1 - \Delta_2.$$

Hence,

$$\underline{\Delta} = \Delta_1 - \Delta_2.$$

Let us apply Hurwicz optimism-pessimism criterion. So, if we apply Hurwicz optimism-pessimism criterion with $\alpha = 0.5$ to the interval

$$[\underline{\Delta}, \overline{\Delta}] = [\Delta_1 - \Delta_2, \Delta_1 + \Delta_2],$$

we end up with the value

$$\Delta = 0.5 \cdot \underline{\Delta} + 0.5 \cdot \overline{\Delta} = \Delta_1.$$

For example, for $\Delta_1 = 10\%$ and $\Delta_2 = 0.1\%$, we get $\Delta = 10\%$—in full accordance with common sense. In other words, *Hurwicz criterion explains the above-described common-sense addition.*

References

1. B. Aryal, L. Bokati, K. Godinez, S. Ibarra, H. Liu, B. Wang, V. Kreinovich, Common sense addition explained by Hurwicz optimism-pessimism criterion, in *Abstracts of the 23rd Joint UTEP/NMSU Workshop on Mathematics, Computer Science, and Computational Sciences* (El Paso, Texas, 3 November 2018)
2. L. Hurwicz, *Optimality Criteria for Decision Making Under Ignorance*, Cowles Commission Discussion Paper, Statistics, No. 370, 1951
3. V. Kreinovich, Decision making under interval uncertainty (and beyond), in *Human-Centric Decision-Making Models for Social Sciences*, ed. by P. Guo, W. Pedrycz (Springer Verlag, 2014), pp.163–193
4. R.D. Luce, R. Raiffa, *Games and Decisions: Introduction and Critical Survey* (Dover, New York, 1989)

Chapter 10
Biased Perception of Time

To make a proper decision, people need to also take into account future consequences of different alternatives. It turns out that they usually underestimate time passed since distant events, and overestimate time passed since recent events. There are several explanations for this "telescoping effect", but most current explanations utilize specific features of human memory and/or human perception. We show that the telescoping effect can be explained on a much basic level of decision theory, without the need to invoke any specific ways we perceive and process time.

Comment. Results from this chapter first appeared in [1].

10.1 Formulation of the Problem

Telescoping effect. It is known that when people estimate how long ago past events happened, their estimates are usually biased (see, e.g., [2–4]):

- for recent events, people usually *overestimate* how much time has passed since this event;
- on the other hand, for events in the more distant past, people usually *underestimate* how much time has passed since the event.

This phenomenon is called *telescoping effect* since the bias in perceiving long-ago past events is similar to what happens when we look at the celestial objects via a telescope: all the objects appear closer than when you look at them with a naked eye.

How can this effect be explained. There are many explanations for the telescoping effect [2–4], but most current explanations utilize specific features of human memory and/or human perception.

L. Bokati and V. Kreinovich, *Decision Making Under Uncertainty, with a Special Emphasis on Geosciences and Education*, Studies in Systems, Decision and Control 218,
https://doi.org/10.1007/978-3-031-26086-5_10

What we do in this chapter. In this chapter, we show that the telescoping effect can be explained on a much basic level of decision theory, without the need to invoke any specific ways we perceive and process time.

10.2 How Decision Theory Can Explain the Telescoping Effect

People's perceptions are imprecise. In the ideal situation, an event of utility u_0 that occurred t moments in the past should be equivalent to exactly the utility $u = q^t \cdot u_0$ now. In practice, however, people's perceptions are imprecise.

Let us describe this imprecision: first approximation. Let us denote by ε the accuracy of people's perception. Then, for an event with actual utility u, the perceived utility can differ by ε, i.e., it can take any value from the corresponding interval $[u - \varepsilon, u + \varepsilon]$. In particular, our perceived utility u of the past event can take any value from the interval $[q^t \cdot u_0 - \varepsilon, q^t \cdot u_0 + \varepsilon]$.

How we perceive events from the distant past. The above interval can be somewhat narrowed down if we take into account that for a positive event, with utility $u_0 > 0$, the perception cannot be negative, while the value $q^t \cdot u_0 - \varepsilon$ is negative for large t. Thus, when $q^t \cdot u_0 - \varepsilon < 0$, i.e., when $t > T_0 \stackrel{\text{def}}{=} \dfrac{\ln(u_0/\varepsilon)}{|\ln(q)|}$, the lower bound of the interval is 0, and thus, the interval has the form

$$[\underline{u}, \overline{u}] = [0, q^t \cdot u_0 + \varepsilon].$$

Based on Hurwicz's optimism-pessimism criterion, this interval is equivalent to the value $\alpha_H \cdot (q^t \cdot u_0 + \varepsilon)$. How does this translate into a perceived time? For any time t_p, the utility of the event t_p moments in the past is equal to $q^{t_p} \cdot u_0$. Thus, the perceived time t_p can be found from the condition that the utility $\alpha_H \cdot (q^t + \varepsilon)$ is equal to $q^{t_p} \cdot u_0$. This equality $\alpha_H \cdot (q^t \cdot u_0 + \varepsilon) = q^{t_p} \cdot u_0$ implies that

$$t_p = \frac{\ln((\alpha_H \cdot (q^t \cdot u_0 + \varepsilon))/u_0)}{\ln(q)}.$$

In particular, when t tends to infinity, we have $q^t \to 0$ and thus, the perceived time tends to a finite constant

$$\frac{\ln((\alpha_H \cdot \varepsilon)/u_0)}{\ln(q)}.$$

Thus, for large t we indeed have $t_p \ll t$, which is exactly what we observe in the telescoping effect for events from the distant past.

How we perceive very recent events. For recent events, the interval

$$[q^t \cdot u_0 - \varepsilon, q^t \cdot u_0 + \varepsilon]$$

can also be somewhat narrowed down if we take into account that the perceived utility of a past event cannot exceed its utility now, i.e., the value u_0. Thus, when $q^t \cdot u_0 + \varepsilon > u_0$, i.e., when $q^t > 1 - \varepsilon/u_0$ and thus, $t < t_0 \stackrel{\text{def}}{=} \dfrac{\ln(1 - u_0/\varepsilon)}{\ln(q)}$, the upper bound of the interval is u_0, and thus, the interval has the form

$$[\underline{u}, \overline{u}] = [q^t \cdot u_0 - \varepsilon, u_0].$$

Based on Hurwicz's optimism-pessimism criterion, this interval is equivalent to the value $\alpha_H \cdot u_0 + (1 - \alpha_H) \cdot (q^t \cdot u_0 - \varepsilon)$. Similarly to the distant-past case, the perceived time t_p can be found from the condition that the above value is equal to $q^{t_p} \cdot u_0$, i.e., that

$$\alpha_H \cdot u_0 + (1 - \alpha_H) \cdot (q^t - \varepsilon) = q^{t_p} \cdot u_0.$$

This implies that

$$t_p = \frac{\ln(\alpha_H + (1 - \alpha_H) \cdot (q^t - \varepsilon/u_0))}{\ln(q)}.$$

In particular, when t tends to 0, we have $q^t \to 1$ and thus, the perceived time t_p tends to a finite positive constant

$$\frac{\ln(\alpha_H + (1 - \alpha_H) \cdot (1 - \varepsilon/u_0))}{\ln(q)}.$$

Thus, for small t, we indeed have $t_p \gg t$, which is exactly what we observe in the telescoping effect for recent events.

References

1. L. Bokati, V. Kreinovich, Decision theory explains 'telescoping effect'-that our time perception is biased. J. Uncertain Syst. **13**(2), 100–103 (2019)
2. S.E. Crawley, L. Pring, When did Mrs. Thatcher resign? The effects of ageing on the dating of public events. Memory **8**(2), 111–121 (2000)
3. S.M.J. Janssen, A.G. Chessa, J.M.J. Murre, Memory for time: how people date events. Mem. Cognit. **34**(1), 138–147 (2006)
4. Wikipedia, *Telescoping effect*, https://en.wikipedia.org/wiki/Telescoping_effect. Downloaded on 29 Dec 2018

Chapter 11
Biased Perception of Future Time Leads to Non-Optimal Decisions

In general, biased perception of time leads decision makers to non-optimal solutions. One of the possible cases of such behavior is the case of temptation. In this chapter, we show that temptation can be explained by decision theory. We hope that this explanation will eventually lead to an accurate prediction of this phenomenon.

Comment. The results of this chapter first appeared in [1].

What is temptation. A popular book [2] by a Nobelist Richard H. Thaler starts the chapter on temptation (Chapter 2) with a simple example: a group of friends are given a big bowl of nuts before dinner. As they eat more and more nuts, they realize that if they continue, they will have no appetite for the incoming tasty dinner, so they decided to put away the bowl.

All this sounds reasonable, until we start analyzing it from the economic viewpoint. From this viewpoint, the more options we have, the better, so how come the elimination of one of the options made everyone happier?

This is just one example; for other examples and for a general analysis of this phenomenon, see, e.g., [2–10].

What if we take discounting into account. Let us try to resolve this puzzle by taking discounting into account. Let us denote the overall amount of food that a person can eat in the evening by a (e.g., by a grams), the utility for eating one gram of nuts by n, the utility of eating one gram of dinner by d, the discounting coefficient from dinner to now by q_+, and the amount of nuts that we eat now by x. The variable x can take any value from the interval $[0, a]$.

In terms of these notations, when we eat x grams of nuts and $a - x$ grams of actual dinner, then, taking into account discounting, the overall utility now is equal to

$$n \cdot x + q_+ \cdot d \cdot (a - x). \tag{11.1}$$

L. Bokati and V. Kreinovich, *Decision Making Under Uncertainty, with a Special Emphasis on Geosciences and Education*, Studies in Systems, Decision and Control 218,
https://doi.org/10.1007/978-3-031-26086-5_11

According to the usual decision making idea, we want to select the amount x for which this utility is the largest. But the expression (11.1) is linear in x, so its largest value on the interval $[0, a]$ is attained at one of the endpoints of this interval, i.e., either for $x = 0$ or for $x = a$. In the first case, we do not eat any nuts at all, in the second case, we only eat nuts and do not eat any dinner. This may be mathematically reasonable, but this is *not* how people behave! How can we explain how people actually behave?

Taking into account that at different moments of time, people have different preferences. In the previous text, we assumed that the only way a person takes into account future events is by discounting. This would make sense if the same person at different moment of time has the same preferences. In reality, people's preferences change. To some extent, the same person at different moments of time is a kind of a different person. So, a proper way to take that into account is to realize that when a person makes decision, he or she needs to find a compromise between his/her today's interests and his/her interests at other moments of time.

This situation is similar to situation of joint decision making, when several people with somewhat different interests try to come up with a group decision–the only difference is that different people can decide not to cooperate at all, while here, "agents" (i.e., the same person at different moments of time) are "joined at the hip"–decisions by one of them affect another one. Thus, to properly describe decision making, we need to view the problem as a problem of group decision making–group decision making by agents representing the same person at different moments of time.

According to decision theory, a group decision of several cooperating agents should be maximizing the product of their utilities. This is known as *Nash's bargaining solution*; see, e.g., [11–13]. So, in our case, a person making a decision should be maximizing the product of the utilities at different moments of time.

Let us show, on the above example, that this indeed helps us avoid the above un-realistic prediction that we should have $x = 0$ or $x = a$.

How this idea help. Let us consider the simplest case of two moments of time: the original moment of time when we are eating (or not eating) nuts, and the future moment of time when we will be eating dinner. In the original moment of time, the utility is described by the formula (11.1). Similarly, at the next moment of time, the utility is described by a formula $q_- \cdot n \cdot x + d \cdot (a - x)$, for an appropriate discounting coefficient q_-. Thus, the correct value x is the one that maximizes the product

$$(n \cdot x + q_+ \cdot d \cdot (a - x)) \cdot (q_- \cdot n \cdot x + d \cdot (a - x)).$$

This function is quadratic, and, in contrast to linear functions, the maximum of a quadratic function on an interval is not necessarily attained at one of the interval's endpoints.

Let us illustrate it on a simplified example where computations are easy: $a = 1$, $n = 1$, $d = 2$, and $q_+ = q_- = 0.25$. In this case, we maximize the function

$$(x + 0.5 \cdot (1 - x)) \cdot (0.25 \cdot x + 2 \cdot (1 - x)) = (0.5 \cdot x + 0.5) \cdot (2 - 1.75 \cdot x).$$

Differentiating this expression with respect to x and equating the derivative to 0 leads to

$$0.5 \cdot (2 - 1.75 \cdot x) + (0.5 \cdot x + 0.5) \cdot (-1.75) = 0,$$

i.e., to $0.125 = 1.75 \cdot x$ and

$$x = \frac{0.125}{1.75} = \frac{1/8}{7/4} = \frac{1}{14} \approx 0.07.$$

The values a, n, etc., were kind of random, but the resulting proportion of nuts snack in the food–about 7%–is quite reasonable.

Comment. So why is everyone happy that the temptation was taken away? Because this allowed everyone not to violate their social contract–in this case, a social contract (as described by Nash's bargaining solution) between a person now and the same person in the future.

References

1. L. Bokati, O. Kosheleva, V. Kreinovich, N.N Thach, Economics of reciprocity and temptation. Technical Report UTEP-CS-20-43 (Department of Computer Science, University of Texas at El Paso, 2020)
2. R.H. Thaler, C.R. Sunstein, *Nudge: Improving Decisions About Health, Wealth, and Happiness* (Penguin Books, New York, 2009)
3. C.F. Camerer, Neuroeconomics: using neuroscience to make economic predictions. Econ. J. **117**, C26–C42 (2007)
4. S. Frederick, G. Loewenstein, T. O'Donoghue, Time discounting and time preference: a critical review. J. Econ. Lit. **40**, 222–226 (2001)
5. J. Gruber, Smoking's 'internalities'. Regul. **25**(4), 52–57 (2002)
6. D. Laibson, Golden eggs and hyperbolic discounting. Q. J. Econ. **112**, 443–477 (1997)
7. S.M. McClure, D.I. Laibson, G. Loewenstein, J.D. Cohen, Separate neural systems value immediate and delayed monetary rewards. Sci. **306**, 503–507 (2004)
8. T. O'Donoghue, M. Rabin, Studying optimal paternalism, illustrated by a model of sin taxes. Am. Econ. Rev. **89**(1), 103–124 (1999)
9. R.H. Thaler, E.J. Johnson, Gambling with the house money and trying to break even: the effects of prior outcomes on risky choice. Manag. Sci. **36**, 643–660 (1990)
10. R.H. Thaler, H.M. Shefrin, An economic theory of self-control. J. Polit. Econ. **89**, 392–406 (1981)
11. R.D. Luce, R. Raiffa, *Games and Decisions: Introduction and Critical Survey* (Dover, New York, 1989)
12. J. Nash, The bargaining problem. Econ. **18**(2), 155–162 (1950)
13. H.P. Nguyen, L. Bokati, V. Kreinovich, New (simplified) derivation of Nash's bargaining solution. J. Adv. Comput. Intell. Intell. Inform. (JACIII). **24**(5), 589–592 (2020)

[illegible]

Let us illustrate [illegible] a simplified example where computations are easy: $a = 1$, [illegible] = [illegible] and [illegible] $= 0$. [illegible] In this case, we standardize the function

$$(a + 0.5 \cdot (1 - x)) \cdot (0.2 \cdot [illegible] + 0.5 \cdot (1 - [illegible])) - 0.35 \cdot x [illegible] (2 - 1/2 [illegible])$$

Differentiating the [illegible] with respect to [illegible] and equating the derivative to 0 leads to [illegible]

$$[illegible] 0.5 [illegible] 2 - 1.75 [illegible] + (0.3 [illegible] + 0.5) [illegible] 1.75 [illegible] = 0$$

[illegible] 1.75.

$$[illegible] = \frac{[illegible]}{[illegible]} = \frac{1.8}{2.4} [illegible] \frac{1.12}{[illegible]} = [illegible]$$

[illegible] ate [illegible] obtained [illegible] the [illegible] proportion [illegible] in the [illegible] out [illegible] some reasoning.

[illegible] that the [illegible] attempt [illegible] Probably [illegible] this [illegible] the [illegible] of [illegible] Nash's [illegible] condition) [illegible] a [illegible] now [illegible] same [illegible]

References

1. [illegible] Texas [illegible]
2. [illegible] [illegible]
3. [illegible] economic predictions [illegible] **117** [illegible]
4. [illegible] Review [illegible] **222** [illegible]
5. [illegible] (2002)
6. D. [illegible] J. Econ. **112** [illegible]
7. [illegible] Sci. **66** [illegible]
8. [illegible] M. Palm [illegible] (1999)
9. R. [illegible] **36** [illegible]
10. R. H. [illegible] **84** [illegible] 106 [illegible]
11. [illegible] New York [illegible]
12. [illegible] Econ. **18**(2) [illegible] (1950)
13. [illegible] **7**(1) [illegible]

Chapter 12
People Have Biased Perception of Other People's Utility

In the idealized description of decision making, people have a perfect knowledge of each other's utility. In practice, however, their perceptions are biased. In this chapter, we provide an explanation for this bias and show that this explains the phenomenon of reciprocity–which seems to contradict the idealized decision making recommendations.

Comment. Results of this chapter first appeared in [1].

What is reciprocity. Usually, people have reasonably fixed attitude to others: they feel empathy towards members of their family, members of their tribe, usually citizens of their country–and may be consistently negative towards their country's competitors. However, in addition to these consistent feelings, they also have widely fluctuating attitudes towards people with whom they work–or at least with whom they are teamed up in a experiment set up by a behavioral economics researcher.

It turns out that while it is difficult to predict how these attitudes will evolve–even in what direction they will evolve, positive or negative–there is a general phenomenon that people are nice to those who treat them nicely and negative to those who treat them badly. In terms of the coefficients α_{ij} it means that:

- if α_{ji} is positive, then we expect α_{ij} to be positive too, and
- if α_{ji} is negative, then we expect α_{ij} to be negative too;

see, e.g., [2, 3].

This *reciprocity* phenomenon is intuitively clear–this is, after all, a natural human behavior–but how can we explain it in economic terms?

L. Bokati and V. Kreinovich, *Decision Making Under Uncertainty, with a Special Emphasis on Geosciences and Education*, Studies in Systems, Decision and Control 218,
https://doi.org/10.1007/978-3-031-26086-5_12

Let us formulate the problem in precise terms. To explain the reciprocity phenomenon, let us consider the simplest case of formula (3.1), when we have only two people. In this case, the formula (3.1) for these two people takes the following form:

$$u_1 = u_1^{(0)} + \alpha_{12} \cdot u_2; \tag{12.1}$$

$$u_2 = u_2^{(0)} + \alpha_{21} \cdot u_1. \tag{12.2}$$

Since each person tries to maximize his/her utility, a natural question is as follows:

- suppose that Person 1 knows the attitude α_{21} of Person 2 towards him/her;
- what value α_{12} describing his/her attitude should Person 1 select to maximize his/her utility u_1?

Analysis of the problem. If we replace, in the right-hand side of the equality (12.1), the value u_2 with the right-hand side of the expression (12.2), we get

$$u_1 = u_1^{(0)} + \alpha_{12} \cdot u_2^{(0)} + \alpha_{12} \cdot \alpha_{21} \cdot u_1.$$

If we move all the terms containing u_1 into the left-hand side, we get

$$u_1 \cdot (1 - \alpha_{12} \cdot \alpha_{21}) = u_1^{(0)} + \alpha_{12} \cdot u_2^{(0)},$$

hence

$$u_1 = \frac{u_1^{(0)} + \alpha_{12} \cdot u_2^{(0)}}{1 - \alpha_{12} \cdot \alpha_{21}}. \tag{12.3}$$

This expression can take infinite value–i.e., as large a value as possible–if we take the value

$$\alpha_{12} = \frac{1}{\alpha_{21}}, \tag{12.4}$$

for which the denominator is 0. We can make it positive–and as large as possible–if we take α_{12} close to the inverse $1/\alpha_{21}$, so that the difference $1 - \alpha_{12} \cdot \alpha_{21}$ will not be exactly 0, but be close to 0, with the same sign as the expression $u_1^{(0)} + \alpha_{12} \cdot u_2^{(0)}$.

This explains reciprocity. Indeed, according to the formula (12.4):

- if α_{21} is positive, then the selected value α_{12} is also positive, and
- if α_{21} is negative, then the selected value α_{12} is also negative.

References

1. L. Bokati, O. Kosheleva, V. Kreinovich, N.N Thach, Economics of reciprocity and temptation. Technical Report UTEP-CS-20-43 (Department of Computer Science, University of Texas at El Paso, 2020)
2. M. Rabin, Incorporating fairness into game theory and economics. Am. Econ. Rev. **83**(5), 1281–1302 (1993)
3. R.H. Thaler, *Misbehaving: The Making of Behavioral Economy* (W.W. Norton & Co., New York, 2015)

Chapter 13
People Select Approximately Optimal Alternatives

According to the ideal decision recommendations, when presented with several choices with different expected equivalent monetary gain, we should select the alternative with the largest gain. In practice, instead, we make a random choice, with probability decreasing with the gain—so that it is possible that we will select second highest and even third highest value. Specifically, we use the so-called softmax formula. Interestingly, the same formula is used in deep learning—and its use increases the learning efficiency.

This formula assumes that we know the exact value of the expected gain for each alternative. In practice, we usually know this gain only with some certainty. For example, often, we only know the lower bound $\underline{f}$ and the upper bound $\overline{f}$ on the expected gain, i.e., we only know that the actual gain f is somewhere in the interval $\left[\underline{f}, \overline{f}\right]$. In this chapter, we show how to extend softmax and the resulting choice formulas to interval uncertainty.

Comment. The results of this chapter first appeared in [1].

13.1 People Use Softmax Instead of Optimization

How people actually make decisions? If a person needs to select between several alternatives $a_1, \ldots, a_n$, and this person knows the exact monetary values $f_1, \ldots, f_n$ associated with each alternative, then we expect this person to always select the alternative with the largest possible monetary value—actual or equivalent. We also expect that if we present the person with the exact same set of alternatives several times in a row, this person will always make the same decision—of selecting the best alternative.

Interestingly, this is *not* how most people make decisions. It turns out that we make decisions probabilistically: instead of always selecting the best alternative, we select each alternative a_i with probability p_i described by the formula

L. Bokati and V. Kreinovich, *Decision Making Under Uncertainty, with a Special Emphasis on Geosciences and Education*, Studies in Systems, Decision and Control 218,
https://doi.org/10.1007/978-3-031-26086-5_13

$$p_i = \frac{\exp(k \cdot f_i)}{\sum\limits_{j=1}^{n} \exp(k \cdot f_j)}, \tag{13.1}$$

for some $k > 0$.

In other words, in most cases, we usually indeed select the alternative with the higher monetary value, but with some probability, we will also select the next highest, with some smaller probability, the next next highest, etc.

This fact was discovered by an economist D. McFadden—who received a Nobel Prize in Economics for this discovery; see, e.g., [2–4].

Why: a qualitative explanation. A reader who is not familiar with numerical methods may expect that if we want to reach the global maximum, we should always select the alternative with the largest estimate of expected gain. This idea was indeed tried in numerical methods—but it does not work well: instead of finding the best alternative, the optimizing algorithm would sometimes get stuck in a local maximum of the corresponding objective function.

In numerical analysis, a usual way to get out of a local minimum is to perform some random change. This is, e.g., the main idea behind simulated annealing. Crudely speaking, it means that we do not always follow the smallest—or the largest—value of the corresponding objective function, we can follow the next smallest (largest), next next smallest, etc.—with some probability.

Similar phenomenon occurs in deep learning. At present, the most efficient machine learning technique is *deep learning* (see, e.g., [5, 6]), in particular, *reinforcement deep learning* [7], where, in addition to processing available information, the system also (if needed) automatically decides which additional information to request—and if an experimental setup is automated, to produce.

For selecting the appropriate piece of information, the system estimates, for each possible alternative, how much information this particular alternative will bring. For the same reason as before, the best optimization result happens when we add randomness.

Softmax: how randomness is currently added. Of course, the actual maximum should be selected with the highest probability, the next value with lower probability, etc. In other words, if we want to maximize some objective function $f(a)$, and we have alternatives $a_1, \ldots, a_n$ for which this function has values $f_1 \stackrel{\text{def}}{=} f(a_1), \ldots, f_n \stackrel{\text{def}}{=} f(a_n)$, then the probability p_i of selecting the i-th alternative should be increasing with f_i, i.e., we should have $p_i \sim F(f_i)$ for some increasing function $F(z)$, i.e., $p_i = c \cdot F(f_i)$, for some constant c.

We should always select one of the alternatives, so these probabilities should add up to 1: $\sum\limits_{j=1}^{n} p_j = 1$. From this condition, we conclude that $c \cdot \sum\limits_{j=1}^{n} F(f_j) = 1$. Thus, $c = 1/\left(\sum\limits_{j=1}^{n} F(f_j)\right)$ and so,

$$p_i = \frac{F(f_i)}{\sum_{j=1}^{n} F(f_j)}. \quad (13.2)$$

Which function $F(z)$ should we choose? In deep learning—a technique that requires so many computations that it cannot exist without high performance computing—computation speed is a must. It is also a must in human decision making, since we often need to make decisions fast, and our computational abilities are much slower than computers'.

Thus, the function $F(z)$ should be fast to compute—which means, in practice, that it should be one of the basic functions for which we have already gained an experience of how to compute it fast. There are a few such functions: arithmetic functions, the power function, trigonometric functions, logarithm, exponential function, etc.

The selected function should be increasing, and it should return non-negative results for all real values z (positive or negative)—otherwise, we will end up with meaningless negative probability. Among basic functions, only one function has this property—the exponential function $F(z) = \exp(k \cdot z)$ for some $k > 0$. For this function, the probability p_i takes the form (13.1). This expression is known as the *softmax* formula.

13.2 Problem: Need to Generalize Softmax to the Case of Interval Uncertainty

When we apply the softmax formula, we only take into account the corresponding estimates $f_1, \ldots, f_n$. However, in practice, we do not just have these estimates, we often have some idea of how accurate is each estimate. Some estimates may be more accurate, some may be less accurate. It is desirable to take this information about uncertainty into account.

For example, we may know the upper bound Δ_i on the absolute value

$$|f_i - f_i^{\text{act}}| \quad (13.3)$$

of the difference between the estimate f_i and the (unknown) actual value f_i^{act} of the objective function. In this case, the only information that we have about the actual value f_i^{act} is that this value is located in the interval $[f_i - \Delta_i, f_i + \Delta_i]$.

How to take this interval information into account when computing the corresponding probabilities p_i? This is the problem that we study in this chapter—and for which we provide a reasonable solution.

13.3 How to Generalize: The Proposed Solution

Discussion. Let $\mathscr{A}$ denote the class of all possible alternatives. We would like, given any finite set of alternatives $A \subseteq \mathscr{A}$ and a specific alternative $a \in A$, to describe the probability $p(a \mid A)$ that out of all the alternatives from the set A, the alternative a will be selected.

Once we know these probabilities, we can then compute, for each set $B \subseteq A$, the probability $p(B \mid A)$ that one of the alternatives from the set B will be selected as $p(B \mid A) = \sum\limits_{b \in B} p(b \mid A)$. In particular, we have $p(a \mid A) = p(\{a\} \mid A)$.

A natural requirement related to these conditional probabilities is that if we have $A \subseteq B \subseteq C$, then we can view the selection of A from C as either a direct selection, or as first selecting B, and then selecting A out of B. The resulting probability should be the same, so we must have $p(A \mid C) = p(A \mid B) \cdot p(B \mid C)$. Thus, we arrive at the following definition.

Definition 13.1 Let $\mathscr{A}$ be a set. Its elements will be called *alternatives*. By a *choice function*, we mean a function $p(a \mid A)$ that assigns to each pair $\langle A, a\rangle$ of a finite set $A \subseteq \mathscr{A}$ and an element $a \in A$ a number from the interval $(0, 1]$ in such a way that the following two conditions are satisfied:

- for every set A, we have $\sum\limits_{a \in A} p(a \mid A) = 1$, and
- whenever $A \subseteq B \subseteq C$, we have $p(A \mid C) = p(A \mid B) \cdot p(B \mid C)$, where

$$p(B \mid A) \stackrel{\text{def}}{=} \sum_{b \in B} p(b \mid A). \tag{13.4}$$

Proposition 13.1 *For each set $\mathscr{A}$, the following two conditions are equivalent to each other:*

- *the function $p(a \mid A)$ is a choice function, and*
- *there exists a function $v : \mathscr{A} \to \mathbb{R}^+$ that assigns a positive number to each alternative $a \in \mathscr{A}$ such that*

$$p(a \mid A) = \frac{v(a)}{\sum\limits_{b \in A} v(b)}. \tag{13.5}$$

Proof It is easy to check that for every function v, the expression (13.5) indeed defines a choice function. So, to complete the proof, it is sufficient to prove that every choice function has the form (13.5). □

Indeed, let $p(a \mid A)$ be a choice function. Let us pick any $a_0 \in \mathscr{A}$, and let us define a function v as

$$v(a) \stackrel{\text{def}}{=} \frac{p(a \mid \{a, a_0\})}{p(a_0 \mid \{a, a_0\})}. \tag{13.6}$$

In particular, for $a = a_0$, both probabilities $p(a \mid \{a, a_0\})$ and $p(a_0 \mid \{a, a_0\})$ are equal to 1, so the ratio $v(a_0)$ is also equal to 1. Let us show that the choice function has the form (13.5) for this function v.

By definition of $v(a)$, for each a, we have $p(a \mid \{a, a_0\}) = v(a) \cdot p(a_0 \mid \{a, a_0\})$.

By definition of a choice function, for each set A containing a_0, we have $p(a \mid A) = p(a \mid \{a, a_0\}) \cdot p(\{a, a_0\} \mid A)$ and $p(a_0 \mid A) = p(a_0 \mid \{a, a_0\}) \cdot p(\{a, a_0\} \mid A)$. Dividing the first equality by the second one, we get

$$\frac{p(a \mid A)}{p(a_0 \mid A)} = \frac{p(a \mid \{a, a_0\})}{p(a_0 \mid \{a, a_0\})}. \tag{13.7}$$

By definition of $v(a)$, this means that

$$\frac{p(a \mid A)}{p(a_0 \mid A)} = v(a). \tag{13.8}$$

Similarly, for each $b \in A$, we have

$$\frac{p(b \mid A)}{p(a_0 \mid A)} = v(b). \tag{13.9}$$

Dividing (13.8) by (13.9), we conclude that for each set A containing a_0, we have

$$\frac{p(a \mid A)}{p(b \mid A)} = \frac{v(a)}{v(b)}. \tag{13.10}$$

Let us now consider a set B that contains a and b but that does not necessarily contain a_0. Then, by definition of a choice function, we have

$$p(a \mid B) = p(a \mid \{a, b\}) \cdot p(\{a, b\} \mid B) \tag{13.11}$$

and

$$p(b \mid B) = p(b \mid \{a, b\}) \cdot p(\{a, b\} \mid B). \tag{13.12}$$

Dividing (13.11) by (13.12), we conclude that

$$\frac{p(a \mid B)}{p(b \mid B)} = \frac{p(a \mid \{a, b\})}{p(b \mid \{a, b\})}. \tag{13.13}$$

The right-hand side of this equality does not depend on the set B. So the left-hand side, i.e., the ratio

$$\frac{p(a \mid B)}{p(b \mid B)} \tag{13.14}$$

also does not depend on the set B. In particular, for the sets B that contain a_0, this ratio—according to the formula (13.10)—is equal to $v(a)/v(b)$. Thus, the same equality (13.10) holds for all sets A—not necessarily containing the element a_0.

From the formula (13.10), we conclude that

$$\frac{p(a \mid A)}{v(a)} = \frac{p(b \mid A)}{v(b)}. \tag{13.15}$$

In other words, for all elements $a \in A$, the ratio

$$\frac{p(a \mid A)}{v(a)} \tag{13.16}$$

is the same. Let us denote this ratio by c_A; then, for each $a \in A$, we have:

$$p(a \mid A) = c_A \cdot v(a). \tag{13.17}$$

From $\sum\limits_{b \in A} p(b \mid A) = 1$, we can now conclude that: $c_A \cdot \sum\limits_{b \in A} v(b) = 1$, thus

$$c_A = \frac{1}{\sum\limits_{b \in A} v(b)}. \tag{13.18}$$

Substituting this expression (13.18) into the formula (13.17), we get the desired expression (13.5).

The proposition is proven.

Comment. This proof is similar to the proofs from [8, 9].

Discussion. As we have mentioned earlier, a choice is rarely a stand-alone event. Usually, we make several choices—and often, at the same time. Let us consider a simple situation. Suppose that we need to make two independent choices:

- in the first choice, we must select one of the alternatives $a_1, \ldots, a_n$, and
- in the second choice, we must select one of the alternatives $b_1, \ldots, b_m$.

We can view this as two separate selection processes. In this case, in the first process, we select each alternative a_i with probability $v(a_i)/\left(\sum_{k=1}^n v(a_k)\right)$ and, in the second process, we select each alternative b_j with probability $v(b_j)/\left(\sum_{\ell=1}^m v(b_\ell)\right)$. Since the two processes are independent, for each pair $\langle a_i, b_j \rangle$, the probability of selecting this pair is equal to the product of the corresponding probabilities:

$$\frac{v(a_i)}{\sum\limits_{k=1}^{n} v(a_k)} \cdot \frac{v(b_j)}{\sum\limits_{\ell=1}^{m} v(b_\ell)}. \tag{13.19}$$

Alternatively, we can view the whole two-stage selection as a single selection process, in which we select a pair $\langle a_i, b_j\rangle$ of alternatives out of all $n \cdot m$ possible pairs. In this case, the probability of selecting a pair is equal to

$$\frac{v(\langle a_i, b_j\rangle)}{\sum_{k=1}^{n}\sum_{\ell=1}^{m} v(\langle a_k, b_\ell\rangle)}. \tag{13.20}$$

The probability of selecting a pair should be the same in both cases, so the values (13.19) and (13.20) must be equal to each other. This equality limits possible functions $v(a)$.

Indeed, if all we know about each alternative a is the interval $\left[\underline{f}(a), \overline{f}(a)\right]$ of possible values of the equivalent monetary gain, then the value v should depend only on this information, i.e., we should have $v(a) = V\left(\underline{f}(a), \overline{f}(a)\right)$ for some function $V(x, y)$. Which functions $V(x, y)$ guarantee the above equality?

To answer this question, let us analyze how the gain corresponding to selecting a pair $\langle a_i, b_j\rangle$ is related to the gains corresponding to individual selections of a_i and b_j. Suppose that for the alternative a_i, our gain $f_i \stackrel{\text{def}}{=} f(a_i)$ can take any value from the interval $\left[\underline{f}_i, \overline{f}_i\right] \stackrel{\text{def}}{=} \left[\underline{f}(a_i), \overline{f}(a_i)\right]$, and for the alternative b_j, our gain $g_j \stackrel{\text{def}}{=} f(b_j)$ can take any value from the interval $\left[\underline{g}_j, \overline{g}_j\right] \stackrel{\text{def}}{=} \left[\underline{f}(b_j), \overline{f}(b_j)\right]$. These selections are assumed to be independent. This means that we can have all possible combinations of values $f_i \in \left[\underline{f}_i, \overline{f}_i\right]$ and $g_j \in \left[\underline{g}_j, \overline{g}_j\right]$.

The smallest possible value of the overall gain $f_i + g_j$ is when both gains are the smallest. In this case, the overall gain is $\underline{f}_i + \underline{g}_j$. The largest possible value of the overall gain $f_i + g_j$ is when both gains are the largest. In this case, the overall gain is $\overline{f}_i + \overline{g}_j$. Thus, the interval of possible values of the overall gain is

$$\left[\underline{f}(\langle a_i, b_j\rangle), \overline{f}(\langle a_i, b_j\rangle)\right] = \left[\underline{f}_i + \underline{g}_j, \overline{f}_i + \overline{g}_j\right]. \tag{13.21}$$

In these terms, the requirement that the expressions (13.19) and (13.20) coincide takes the following form:

Definition 13.2 We say that a function $V : \mathbb{R} \times \mathbb{R} \to \mathbb{R}^+$ is *consistent* if for every two sequences of intervals $\left[\underline{f}_1, \overline{f}_1\right], \ldots, \left[\underline{f}_n, \overline{f}_n\right]$, and $\left[\underline{g}_1, \overline{g}_1\right], \ldots, \left[\underline{g}_m, \overline{g}_m\right]$, for every i and j, we have

$$\frac{V\left(\underline{f}_i, \overline{f}_i\right)}{\sum_{k=1}^{n} V\left(\underline{f}_k, \overline{f}_k\right)} \cdot \frac{V\left(\underline{g}_j, \overline{g}_j\right)}{\sum_{\ell=1}^{m} V\left(\underline{g}_\ell, \overline{g}_\ell\right)} = \frac{V\left(\underline{f}_i + \underline{g}_j, \overline{f}_i + \overline{g}_j\right)}{\sum_{k=1}^{n}\sum_{\ell=1}^{m} V\left(\underline{f}_k + \underline{g}_\ell, \overline{f}_k + \overline{g}_\ell\right)}. \tag{13.22}$$

Monotonicity. Another reasonable requirement is that the larger the expected gain, the more probable that the corresponding alternative is selected.

The notion of "larger" is easy when gains are exact, but for intervals, we can provide the following definition.

Definition 13.3 We say that an interval A is *smaller than or equal to* an interval B (and denote it by $A \le B$) if the following two conditions hold:

- for every element $a \in A$, there is an element $b \in B$ for which $a \le b$, and
- for every element $b \in B$, there is an element $a \in A$ for which $a \le b$.

Proposition 13.2 $[\underline{a}, \overline{a}] \le [\underline{b}, \overline{b}] \Leftrightarrow (\underline{a} \le \underline{b} \,\&\, \overline{a} \le \overline{b})$.

Proof is straightforward. □

Definition 13.4 We say that a function $V : \mathbb{R} \times \mathbb{R} \to \mathbb{R}^+$ is *monotonic* if for every two intervals $[\underline{a}, \overline{a}]$ and $[\underline{b}, \overline{b}]$, if $[\underline{a}, \overline{a}] \le [\underline{b}, \overline{b}]$ then $V(\underline{a}, \overline{a}) \le V(\underline{b}, \overline{b})$.

Proposition 13.3 *For each function $V : \mathbb{R} \times \mathbb{R} \to \mathbb{R}^+$, the following two conditions are equivalent to each other:*

- *the function V is consistent and monotonic;*
- *the function $V(\underline{f}, \overline{f})$ has the form*

$$V(\underline{f}, \overline{f}) = C \cdot \exp\left(k \cdot \left(\alpha_H \cdot \overline{f} + (1 - \alpha_H) \cdot \underline{f}\right)\right) \tag{13.23}$$

for some values $C > 0$, $k > 0$, and $\alpha_H \in [0, 1]$.

Conclusion. Thus, if we have n alternatives $a_1, \ldots, a_n$, and for each alternative a_i, we know the interval $[\underline{f}_i, \overline{f}_i]$ of possible values of the gain, we should select each alternative i with the probability

$$p_i = \frac{\exp\left(k \cdot \left(\alpha_H \cdot \overline{f}_i + (1 - \alpha_H) \cdot \underline{f}_i\right)\right)}{\sum_{j=1}^{n} \exp\left(k \cdot \left(\alpha_H \cdot \overline{f}_j + (1 - \alpha_H) \cdot \underline{f}_j\right)\right)}. \tag{13.24}$$

So, *we have extended the softmax/McFadden's discrete choice formula to the case of interval uncertainty.*

Comment 1. Proposition 13.3 justifies the formula (13.24). It should be mentioned that the formula (13.24) coincides with what we would have obtained from the original McFadden's formula if, instead of the exact gain f_i, we substitute into this original formula, the expression $f_i = \alpha_H \cdot \overline{f}_i + (1 - \alpha_H) \cdot \underline{f}_i$ for some $\alpha_H \in [0, 1]$. This expression was first proposed by a future Nobelist Leo Hurwicz and is thus known as Hurwicz optimism-pessimism criterion [10–13].

Comment 2. For the case when we know the exact values of the gain, i.e., when we have a degenerate interval $[f, f]$, we get a *new justification for the original McFadden's formula.*

Comment 3. Similar ideas can be used to extend softmax and McFadden's formula to other types of uncertainty. As one can see from the proof, by taking logarithm of V, we reduce the consistency condition to additivity, and additive functions are known; see, e.g., [12]. For example, for probabilities, the equivalent gain is the expected value—since, due to large numbers theorem, the sum of many independent copies of a random variable is practically a deterministic number. Similarly, a class of probability distributions is equivalent to the interval of mean values corresponding to different distributions, and specific formulas are known for the fuzzy case.

Proof of Proposition 13.3. It is easy to check that the function (13.24) is consistent and monotonic. So, to complete the proof, it is sufficient to prove that every consistent monotonic function has the desired form.

Indeed, let us assume that the function V is consistent and monotonic. Then, due to consistency, it satisfies the formula (13.22). Taking logarithm of both sides of the formula (13.22), we conclude that for the auxiliary function $u(\underline{a}, \overline{a}) \stackrel{\text{def}}{=} \ln(V(\underline{a}, \overline{a}))$, for every two intervals $[\underline{a}, \overline{a}]$ and $[\underline{b}, \overline{b}]$, we have

$$u(\underline{a}, \overline{a}) + u\left(\underline{b}, \overline{b}\right) = u\left(\underline{a} + \underline{b}, \overline{a} + \overline{b}\right) + c \tag{13.25}$$

for an appropriate constant c. Thus, for $U(\underline{a}, \overline{a}) \stackrel{\text{def}}{=} u(\underline{a}, \overline{a}) - c$, substituting $u(\underline{a}, \overline{a}) = U(\underline{a}, \overline{a}) + c$ into the formula (13.25), we conclude that

$$U(\underline{a}, \overline{a}) + U\left(\underline{b}, \overline{b}\right) = U\left(\underline{a} + \underline{b}, \overline{a} + \overline{b}\right), \tag{13.26}$$

i.e., that the function U is additive. Similarly to [12], we can use the general classification of additive locally bounded functions (and every monotonic function is locally bounded) from [14] to conclude that $U(\underline{a}, \overline{a}) = k_1 \cdot \overline{a} + k_2 \cdot \underline{a}$. Monotonicity with respect to changes in $\underline{a}$ and $\overline{a}$ imply that $k_1 \geq 0$ and $k_2 \geq 0$. Thus, for

$$V(\underline{a}, \overline{a}) = \exp(u(\underline{a}, \overline{a})) = \exp(U(\underline{a}, \overline{a}) + c) = \exp(c) \cdot \exp(U(\underline{a}, \overline{a})), \tag{13.27}$$

we get the desired formula, with $C = \exp(c)$, $k = k_1 + k_2$ and $\alpha_H = k_1/(k_1 + k_2)$.

The proposition is proven.

This may lead to a further improvement of deep learning. Currently, one of the most promising Artificial Intelligence techniques is deep learning. The successes of using deep learning are spectacular—from winning over human champions in Go (a very complex game that until recently resisted computer efforts) to efficient algorithms for self-driving cars. All these successes require a large amount of computations on high performance computers.

While deep learning has been very successful, there is a lot of room for improvement. For example, the existing deep learning algorithms implicitly assume that all the input data are exact, while in reality, data comes from measurements and measurement are never absolutely accurate. The simplest situation is when we know the upper bound Δ on the measurement error. In this case, based on the measurement result $\widetilde{x}$, the only thing that we can conclude about the actual value x is that x is in the interval $[\widetilde{x} - \Delta, \widetilde{x} + \Delta]$. In this chapter, we have shown how computing softmax—one of the important steps in deep learning algorithms—can be naturally extended to the case of such interval uncertainty. The resulting formulas are almost as simple as the original ones, so their implementation will take about the same time on the same high performance computers.

References

1. B.J. Kubica, L. Bokati, O. Kosheleva, V. Kreinovich, Softmax and McFadden's discrete choice under interval (and other) uncertainty, in *Proceedings of the International Conference on Parallel Processing and Applied Mathematics PPAM'2019*, Bialystok, Poland, September 8–11, 2019, vol. II, ed. by R. Wyrzykowski, E. Deelman, J. Dongarra, and K. Karczewski (Springer, 2020), pp. 364–373
2. D. McFadden, Conditional logit analysis of qualitative choice behavior, in *Frontiers in Econometrics*. ed. by P. Zarembka (Academic, New York, 1974), pp.105–142
3. D. McFadden, Economic choices. Am. Econ. Rev. **91**, 351–378 (2001)
4. K. Train, *Discrete Choice Methods with Simulation* (Cambridge University Press, Cambridge, 2003)
5. I. Goodfellow, Y. Bengio, A. Courville, *Deep Learning* (MIT Press, Cambridge, 2016)
6. V. Kreinovich, From traditional neural networks to deep learning: towards mathematical foundations of empirical successes, in *Proceedings of the World Conference on Soft Computing, Baku, Azerbaijan, May 29–31, 2018*, ed. by S.N. Shahbazova (2018)
7. R.S. Sutton, A.G. Barto, *Reinforcement Learning. An Introduction*, 2nd edn. (MIT Press, Cambridge, 2018)
8. O. Kosheleva, V. Kreinovich, S. Sriboonchitta, Econometric models of probabilistic choice: beyond McFadden's formulas, in *Robustness in Econometrics*, ed. by V. Kreinovich, S. Sriboonchitta, V.N. Huynh (Springer, Cham, 2017), pp. 79–88
9. D. Luce, *Individual Choice Behavior: A Theoretical Analysis* (Dover, New York, 2005)
10. L. Hurwicz, *Optimality Criteria for Decision Making Under Ignorance*, Cowles Commission Discussion Paper, Statistics, No. 370 (1951)
11. V. Kreinovich, Decision making under interval uncertainty (and beyond), in *Human-Centric Decision-Making Models for Social Sciences*. ed. by P. Guo, W. Pedrycz (Springer, Berlin, 2014), pp.163–193
12. V. Kreinovich, Decision making under interval (and more general) uncertainty: monetary versus utility approaches. J. Comput. Technol. **22**(2), 37–49 (2017)
13. R.D. Luce, R. Raiffa, *Games and Decisions: Introduction and Critical Survey* (Dover, New York, 1989)
14. J. Aczél, J. Dhombres, *Functional Equations in Several Variables* (Cambridge University Press, 2008)

Chapter 14
People Make Decisions Using Heuristics. I

In some cases, instead of looking for an optimal solution, people use some heuristic ideas–with the hope that these ideas will lead to reasonable quality decisions. In this chapter and in the following chapter, we will give two examples of such heuristic ideas, and we will provide a justification for these ideas.

In particular, in this section, we analyze the heuristic ideas of using the Maximum Entropy approach for selecting an investment portfolio. The traditional Markowitz approach to portfolio optimization assumes that we know the means, variances, and covariances of the return rates of all the financial instruments. In some practical situations, however, we do not have enough information to determine the variances and covariances, we only know the means. To provide a reasonable portfolio allocation for such cases, researchers proposed a heuristic maximum entropy approach. In this chapter, we provide an economic justification for this heuristic idea.

Comment. Results from this chapter first appeared in [1].

14.1 Formulation of the Problem

Portfolio optimization: general problem. What is the best way to invest money? Usually, there are several possible financial instruments; let us denote the number of available financial instruments by n. The questions is then: what portion w_i of the overall money amount should we allocate to each instrument i? Of course, these portions must be non-negative and add up to one:

$$\sum_{i=1}^{n} w_i = 1. \tag{14.1}$$

L. Bokati and V. Kreinovich, *Decision Making Under Uncertainty, with a Special Emphasis on Geosciences and Education*, Studies in Systems, Decision and Control 218,
https://doi.org/10.1007/978-3-031-26086-5_14

The corresponding tuple $w = (w_1, \ldots, w_n)$ is known as an *investment portfolio*, or simply *portfolio*, for short.

Case of complete knowledge: Markowitz solution. If we place money in a bank, we get a guaranteed interest, with a given rate of return r. However, for most other financial instruments i, the rate of return r_i is not fixed, it changes (e.g., fluctuates) year after year. For each values of instrument returns, the corresponding portfolio return r is equal to $r = \sum\limits_{i=1}^{n} w_i \cdot r_i$.

In many practical situations, we know, from experience, the probabilistic distributions of the corresponding rates of return. Based on this past experience, for each instrument i, we can estimate the expected rate of return $\mu_i = E[r_i]$ and the corresponding standard deviation $\sigma_i = \sqrt{E[(r_i - \mu_i)^2]}$. We can also estimate, for each pair of financial instruments i and j, the covariance

$$c_{ij} \stackrel{\text{def}}{=} E[(r_i - \mu_i) \cdot (r_j - \mu_j)].$$

By using this information, for each possible portfolio $w = (w_1, \ldots, w_n)$, we can compute the expected return

$$\mu = E[r] = \sum_{i=1}^{n} w_i \cdot \mu_i \tag{14.2}$$

and the corresponding variance

$$\sigma^2 = \sum_{i=1}^{n} w_i^2 \cdot \sigma_i^2 + \sum_{i=1}^{n} \sum_{j=1}^{n} c_{ij} \cdot w_i \cdot w_j. \tag{14.3}$$

The larger the expected rate of return μ we want, the largest the risk that we have to take, and thus, the larger the variance. It is therefore reasonable, given the desired expected rate of return μ, to find the portfolio that minimizes the variance, i.e., that minimizes the expression (14.3) under the constraints (14.1) and (14.2).

This problem was first considered by the future Nobelist Markowitz, who proposed an explicit solution to this problem; see, e.g., [2]. Namely, the Lagrange multiplier method enables to reduce this constraint optimization problem to the following unconstrained optimization problem: minimize the expression

$$\sum_{i=1}^{n} w_i^2 \cdot \sigma_i^2 + \sum_{i=1}^{n} \sum_{j=1}^{n} c_{ij} \cdot w_i \cdot w_j + \lambda_1 \cdot \left(\sum_{i=1}^{n} w_i - 1 \right) +$$

$$\lambda_2 \cdot \left(\sum_{i=1}^{n} w_i \cdot \mu_i - \mu \right), \tag{14.4}$$

where λ_1 and λ_2 are Lagrange multipliers that need to be determined from the conditions (14.1) and (14.2).

Differentiating the expression (14.4) by the unknowns w_i, we get the following system of linear equations:

$$2\sigma_i \cdot w_i + 2\sum_{j\neq i} c_{ij} \cdot w_j + \lambda_1 + \lambda_2 \cdot \mu_i = 0. \tag{14.5}$$

Thus,

$$w_i = \lambda_1 \cdot w_i^{(1)} + \lambda_2 \cdot w_i^{(2)}, \tag{14.6}$$

where $w_i^{(j)}$ are solutions to the following systems of linear equations

$$2\sigma_i \cdot w_i + 2\sum_{j\neq i} c_{ij} \cdot w_j = -1 \tag{14.7}$$

and

$$2\sigma_i \cdot w_i + 2\sum_{j\neq i} c_{ij} \cdot w_j = -\mu_i. \tag{14.8}$$

Substituting the expression (14.6) into the Eqs. (14.1) and (14.2), we get a system two linear equations for two unknowns λ_1 and λ_2. From this system, we can easily find the coefficients λ_i and thus, the desired portfolio (14.6).

Case of complete information: modifications of Markowitz solution. Some researchers argue that variance may be not the best way to describe the intuitive notion of risk. Instead, they propose to use other statistical characteristics, e.g., the quantile q_α corresponding to a certain small probability α–i.e., a value for which the probability that the returns are very low ($r \le q_\alpha$) is equal to α.

Instead of the original Markowitz problem, we thus have a problem of maximizing q_α–or another characteristic–under the given expected return μ. Computationally, the resulting constraint optimization problems are no longer quadratic and thus, more complex to solve, but they are still well formulated and thus, solvable.

Case of partial information: formulation of the general problem. In many practical situations, we only have partial information about the probabilities of different rates of return r_i.

For example, in some cases, we know the expected returns μ_i, but we do not have any information about the standard deviations and covariances. What portfolio should we select in such situations?

Maximum Entropy approach: reminder. Situations in which we only have partial information about the probabilities–and thus, several different probability distributions are consistent with the available information–such situations are ubiquitous.

Usually, some of the consistent distributions are more precise, some are more uncertain. We do not want to pretend that we know more than we actually do, so in such situations of uncertainty, a natural idea is to select a distribution which has the largest possible degree of uncertainty. A reasonable way to describe the uncertainty of a probability distribution with the probability density $\rho(x)$ is by its *entropy*

$$S = -\int \rho(x) \cdot \ln(\rho(x))\,dx. \tag{14.9}$$

So, we select the distribution whose entropy is the largest; see, e.g., [3].

In many cases, this *Maximum Entropy* approach makes perfect sense. For example, if the only information that we have about a probability distribution is that it is located on an interval $[\underline{x}, \overline{x}]$, then out of all possible distributions, the Maximum Entropy approach selects the uniform distribution $\rho(x) = \text{const}$ on this interval. This makes perfect sense–if we do not have any reason to believe that one of the values from the interval is more probable than other values, then it makes sense to assume that all the values from this interval are equally probable, which is exactly $\rho(x) = \text{const}$.

In situations when we know marginal distributions of each of the variables, but we do not have any information about the dependence between these variables, the Maximum Entropy approach concludes that these variables are independent. This also makes perfect sense: if we have no reason to believe that the variables are positively or negatively correlated, it makes sense to assume that they are not correlated at all.

If all we know is the mean and the standard deviation, then the Maximum Entropy approach leads to the normal (Gaussian) distribution–which is in good accordance with the fact that such distributions are indeed ubiquitous.

So, in situations when we only have a partial information about the probabilities of different return values, it makes sense to select, out of all possible probability distributions, the one with the largest entropy, and then use this selected distribution to find the corresponding portfolio.

Problem: Maximum Entropy approach is not applicable to the case when we only know μ_i. In many practical situations, the Maximum Entropy approach leads to reasonable results. However, it is not applicable to the situation when we only know the expected rates of return μ_i.

This impossibility can be illustrated already on the case when we have a single financial instrument. Its rate of return r_1 can take any value, positive or negative, the only information that we have about the corresponding probability distribution $\rho(x)$ is that

$$\mu_1 = \int x \cdot \rho(x)\,dx \tag{14.10}$$

and, of course, that $\rho(x)$ is a probability distribution, i.e., that

$$\int \rho(x)\,dx = 1. \tag{14.11}$$

The constraint optimization problem of maximizing the entropy (14.9) under the constraints (14.10) and (14.11) can be reduced to the following unconstrained optimization problem: maximize

$$-\int \rho(x) \cdot \ln(\rho(x))\,dx + \lambda_1 \cdot \left(\int x \cdot \rho(x)\,dx - \mu_1\right) +$$

$$\lambda_2 \cdot \left(\int \rho(x)\,dx - 1\right). \tag{14.12}$$

Differentiating the expression (14.12) with respect to the unknown $\rho(x)$ and equating the derivative to 0, we get

$$-\ln(\rho(x)) - 1 + \lambda_1 \cdot x + \lambda_2 = 0,$$

hence

$$\ln(\rho(x)) = (\lambda_2 - 1) + \lambda_1 \cdot x$$

and $\rho(x) = C \cdot \exp(\lambda_1 \cdot x)$, where $C = \exp(\lambda_2 - 1)$. The problem is that the integral of this exponential function over the real line is always infinite, we cannot get it to be equal to 1–which means that it is not possible to attain the maximum, entropy can be as large as we want.

So how do we select a portfolio in such a situation?

A heuristic idea. In the situation in which we only know the means μ_i, we cannot use the Maximum Entropy approach to find the most appropriate probability distribution. However, here, the portions w_i–since they add up to 1–can also be viewed as kind of probabilities. It therefore makes sense to look for a portfolio for which the corresponding entropy

$$-\sum_{i=1}^{n} w_i \cdot \ln(w_i) \tag{14.13}$$

attains the largest possible value under the constraints (14.1) and (14.2); see, e.g., [4–9].

This heuristic idea sometimes leads to reasonable results. Here, entropy can be viewed as a measure of diversity. Thus, the idea to bring more diversity to one's portfolio makes perfect sense. However, there is a problem.

Remaining problem. The problem is that while the weights w_i do add up to one, they are *not* probabilities. So, in contrast to the probabilistic case, where the Maximum Entropy approach has many justifications, for the weights, there does not seem to be any reasonable justification. It is therefore desirable to either justify this heuristic method–or provide a justified alternative.

What we do in this chapter. In this chapter, we provide a justification for the Maximum Entropy approach. We also show that a similar idea can be applied to a slightly more complex–and more realistic–case, when we only know bounds $\underline{\mu}_i$ and $\overline{\mu}_i$ on the values μ_i.

14.2 Case When We Only Know the Expected Rates of Return μ_i: Economic Justification of the Maximum Entropy Approach

General definition. We want, given n expected return rates $\mu_1, \ldots, \mu_n$, to generate the weights $w_1 = f_{n1}(\mu_1, \ldots, \mu_n)$, ..., $w_n = f_{nn}(\mu_1, \ldots, \mu_n)$ depending on μ_i for which the sum of the weights is equal to 1.

Definition 14.1 By a *portfolio allocation scheme*, we mean a family of functions $f_{ni}(\mu_1, \ldots, \mu_n) \neq 0$ of non-negative variables μ_i, where n is arbitrary integer larger than 1, and $i = 1, 2, \ldots, n$, such that for all n and for all $\mu_i \geq 0$, we have

$$\sum_{i=1}^{n} f_{ni}(\mu_1, \ldots, \mu_n) = 1.$$

Symmetry. Of course, the portfolio allocation should not depend on the order in which we list the instrument.

Definition 14.2 We say that a portfolio allocation scheme is *symmetric* if for each n, for each $\mu_1, \ldots, \mu_n$, for each $i \leq n$, and for each permutation $\pi : \{1, \ldots, n\} \to \{1, \ldots, n\}$, we have

$$f_{ni}(\mu_1, \ldots, \mu_n) = f_{n,\pi(i)}(\mu_{\pi(1)}, \ldots, \mu_{\pi(n)}).$$

Pairwise comparison. If we only have two financial instruments ($n = 2$) with expected rates μ_1 and μ_2, then we assign weights w_1 and $w_2 = 1 - w_1$ depending on the known values μ_1 and μ_2: $w_1 = f_{21}(\mu_1, \mu_2)$ and $w_2 = f_{14}(\mu_1, \mu_2)$.

In the general case, if we have n instruments including these two, then the amount $f_{n1}(\mu_1, \ldots, \mu_n) + f_{n2}(\mu_1, \ldots, \mu_n)$ is allocated for these two instruments. Once this amount is decided on, we should divide it optimally between these two instruments. The optimal division means that the first instrument gets the portion $f_{21}(w_1, w_2)$ of this overall amount, so we must have

$$f_{n1}(\mu_1, \mu_2, \ldots) = f_{21}(\mu_1, \mu_2) \cdot (f_{n1}(\mu_1, \ldots, \mu_n) + f_{n2}(\mu_1, \ldots, \mu_n)). \quad (14.14)$$

Thus, we arrive at the following definition.

Definition 14.3 We say that a portfolio allocation scheme is *consistent* if for every $n > 2$ and for all $i \neq j$, we have

$$f_{ni}(\mu_1, \ldots, \mu_n) = f_{21}(\mu_i, \mu_j) \cdot (f_{ni}(\mu_1, \ldots, \mu_n) + f_{nj}(\mu_1, \ldots, \mu_n)). \quad (14.15)$$

Proposition 14.1 *A portfolio allocation scheme is symmetric and consistent if and only if there exists a function* $f(\mu) \geq 0$ *for which*

$$f_{ni}(\mu_1, \ldots, \mu_n) = \frac{f(\mu_i)}{\sum_{j=1}^{n} f(\mu_j)}. \quad (14.16)$$

Proof It is easy to check that the formula (14.16) describes a symmetric and consistent portfolio allocation scheme. So, to complete the proof, it is sufficient to show that every symmetric and consistent portfolio allocation scheme has the form (14.16).

Indeed, let us assume that the portfolio allocation scheme satisfies the formula (14.15). If we write the formulas (14.15) for i and j and then divide the i-formula by the j-formula, we get the following equality:

$$\frac{f_{ni}(\mu_1, \ldots, \mu_n)}{f_{nj}(\mu_1, \ldots, \mu_n)} = \Phi(\mu_i, \mu_j) \stackrel{\text{def}}{=} \frac{f_{21}(\mu_i, \mu_j)}{f_{21}(\mu_j, \mu_i)}. \quad (14.17)$$

Due to symmetry, $f_{21}(\mu_i, \mu_j) = f_{21}(\mu_j, \mu_i)$, so we have

$$\Phi(\mu_i, \mu_j) = \frac{f_{21}(\mu_i, \mu_j)}{f_{21}(\mu_j, \mu_i)} \quad (14.18)$$

and

$$\Phi(\mu_j, \mu_i) = \frac{f_{21}(\mu_j, \mu_i)}{f_{21}(\mu_i, \mu_j)}, \quad (14.19)$$

thus

$$\Phi(\mu_j, \mu_i) = \frac{1}{\Phi(\mu_i, \mu_j)}. \quad (14.20)$$

Now, for each i, j, and k, we have

$$\frac{f_{ni}(\mu_1, \ldots, \mu_n)}{f_{nj}(\mu_1, \ldots, \mu_n)} = \frac{f_{ni}(\mu_1, \ldots, \mu_n)}{f_{nk}(\mu_1, \ldots, \mu_n)} \cdot \frac{f_{nk}(\mu_1, \ldots, \mu_n)}{f_{nj}(\mu_1, \ldots, \mu_n)},$$

thus

$$\Phi(\mu_i, \mu_j) = \Phi(\mu_i, \mu_k) \cdot \Phi(\mu_k, \mu_j).$$

In particular, for $\mu_k = 1$, we have

$$\Phi(\mu_i, \mu_j) = \Phi(\mu_i, 1) \cdot \Phi(1, \mu_j). \quad (14.21)$$

Due to (14.20), this means that

$$\Phi(\mu_i, \mu_j) = \frac{\Phi(\mu_i, 1)}{\Phi(\mu_j, 1)}, \quad (14.22)$$

i.e.,

$$\Phi(\mu_i, \mu_j) = \frac{f(\mu_i)}{f(\mu_j)}, \quad (14.23)$$

where we denoted $f(\mu) \stackrel{\text{def}}{=} F(\mu, 1)$. Substituting this expression (14.23) into the formula (14.17) and taking $j = 1$, we conclude that

$$\frac{f_{ni}(\mu_1, \ldots, \mu_n)}{f_{n1}(\mu_1, \ldots, \mu_n)} = \frac{f(\mu_i)}{f(\mu_1)}, \quad (14.24)$$

i.e.,

$$f_{ni}(\mu_1, \ldots, \mu_n) = C \cdot f(\mu_i), \quad (14.25)$$

where we denoted

$$C \stackrel{\text{def}}{=} \frac{f_{n1}(\mu_1, \ldots, \mu_n)}{f(\mu_1)}.$$

From the condition that the values f_{nj} corresponding to $j = 1, \ldots, n$ should add up to 1, we conclude that $C \cdot \sum\limits_{j=1}^{n} f(\mu_j) = 1$, hence

$$C = \frac{1}{\sum\limits_{j=1} f(\mu_j)}$$

and thus, the expression (14.25) takes exactly the desired form. The proposition is proven.

Monotonicity. If all we know about each financial instruments is their expected rate of return, then it is reasonable to assume that the larger the expected rate of return, the better the instrument. It is therefore reasonable to require that the larger the rate of return, the larger portion of the original amount should be invested in this instrument.

Definition 14.4 We say that a portfolio allocation scheme is *monotonic* if for each n and each μ_i, if $\mu_i \geq \mu_j$, then $f_{ni}(\mu_1, \ldots, \mu_n) \geq f_{nj}(\mu_1, \ldots, \mu_n)$.

One can easily check that a symmetric and consistent portfolio allocation scheme is monotonic if and only if the corresponding function $f(\mu)$ is non-decreasing.

Shift-invariance. Suppose that, in addition to the return from the investment, a person also get some additional fixed income, which when divided by the amount of money to be invested, translates into the rate r_0. This situation can be described in two different ways:

- we can consider r_0 separately from the investment; in this case, we should allocate, to each financial instrument i, the portion $f_i(\mu_1, \ldots, \mu_n)$;
- alternatively, we can combine both incomes into one and say that for each instrument i, we will get the expected rate of return $\mu_i + r_0$; in this case, to each financial instrument i, we allocate a portion $f_i(\mu_1 + r_0, \ldots, \mu_n + r_0)$.

Clearly, this is the same situation described in two different ways, so the portfolio allocation should not depend on how exactly we represent the same situation. Thus, we arrive at the following definition.

Definition 14.5 We say that a portfolio allocation scheme is *shift-invariant* if for all n, for all $\mu_1, \ldots, \mu_n$, for all i, and for all r_0, we have

$$f_{ni}(\mu_1, \ldots, \mu_n) = f_{ni}(\mu_1 + r_0, \ldots, \mu_n + r_0).$$

Proposition 14.2 *For each portfolio allocation scheme, the following two conditions are equivalent to each other:*

- *the scheme is symmetric, consistent, monotonic, and shift-invariant, and*
- *the scheme has the form*

$$f_{ni}(\mu_1, \ldots, \mu_n) = \frac{\exp(\beta \cdot \mu_i)}{\sum_{j=1}^{n} \exp(\beta \cdot \mu_j)}. \tag{14.26}$$

for some $\beta \geq 0$.

Proof It is clear that the scheme (14.26) has all the desired properties. Vice versa, let us assume that a scheme has all the desired properties. Then, from shift-invariance, for each i and j, we get

$$\frac{f_{ni}(\mu_1, \ldots, \mu_n)}{f_{nj}(\mu_1, \ldots, \mu_n)} = \frac{f_{ni}(\mu_1 + r_0, \ldots, \mu_n + r_0)}{f_{nj}(\mu_1 + r_0, \ldots, \mu_n + r_0)}. \tag{14.27}$$

Substituting the formula (14.16), we conclude that

$$\frac{f(\mu_i)}{f(\mu_j)} = \frac{f(\mu_i + r_0)}{f(\mu_j + r_0)}, \tag{14.28}$$

which implies that

$$\frac{f(\mu_i + r_0)}{f(\mu_i)} = \frac{f(\mu_j + r_0)}{f(\mu_j)}. \tag{14.29}$$

The left-hand side of this equality does not depend on μ_j, the right-hand side does not depend on μ_i. Thus, the ratio depends only on r_0. Let us denote this ratio by $R(r_0)$. Then, we get $f(\mu + r_0) = R(r_0) \cdot f(\mu)$.

It is known (see, e.g., [10]) that every non-decreasing solution to this functional equation has the form $\text{const} \cdot \exp(\beta \cdot \mu)$ for some $\beta \geq 0$. The proposition is proven.

Main result. Now, we are ready to formulate our main result–an economic justification of the above heuristic method.

Proposition 14.3 *Let μ be the desired expected return rate, and assume that we only consider allocation schemes providing this expected return rate, i.e., schemes for which*

$$\sum_{i=1}^{n} \mu_i \cdot w_i = \sum_{i=1}^{n} \mu_i \cdot f_{ni}(\mu_1, \ldots, \mu_n) = \mu. \tag{14.30}$$

Then, the following two conditions on a portfolio allocation schemes are equivalent to each other:

- *the scheme is symmetric, consistent, monotonic, and shift-invariant, and*
- *the scheme has the largest possible entropy $-\sum_{i=1}^{n} w_i \cdot \ln(w_i)$ among all the schemes with the given expected return rate.*

Proof Maximizing entropy under the constraints $\sum w_i \cdot \mu_i = \mu_0$ and $\sum w_i = 1$ is, due to Lagrange multiplier method, equivalent to maximizing the expression

$$-\sum_{i=1}^{n} w_i \cdot \ln(w_i) + \lambda_1 \cdot \left(\sum_{i=1}^{n} w_i \cdot \mu_i - \mu\right) + \lambda_2 \cdot \left(\sum_{i=1}^{n} w_i - 1\right). \tag{14.31}$$

Differentiating this expression by w_i and equating the derivative to 0, we conclude that

$$-\ln(w_i) - 1 + \lambda_1 \cdot \mu_1 + \lambda_2 = 0, \tag{14.32}$$

i.e., that

$$w_i = \text{const} \cdot \exp(\lambda_1 \cdot \mu_i).$$

This is exactly the expression (14.26) which, as we have proved in Proposition 14.2, is indeed equivalent to symmetry, consistency, monotonicity, and shift-invariance. The proposition is proven.

Discussion. What we proved, in effect, is that maximizing diversity is a great idea, be it diversity when distributing money between financial instrument, or–when the state invests in its citizens–when we allocate the budget between cities, between districts, between ethnic groups, or when a company is investing in its future by hiring people of different backgrounds.

14.3 Case When We Only Know the Intervals Containing the Actual (Unknown) Expected Return Rates

Description of the case. Let us now consider an even more realistic case, when we take into account that the expected rates of return μ_i are only approximately known. To be precise, we assume that for each i, we only know the interval $[\underline{\mu}_i, \overline{\mu}_i]$ containing the actual (unknown) expected return rates μ_i. How should we then distribute the investments?

Definition 14.6 By an *interval-based portfolio allocation scheme*, we mean a family of functions $f_{ni}(\underline{\mu}_1, \overline{\mu}_1 \ldots, \underline{\mu}_n, \overline{\mu}_n) \neq 0$ of non-negative variables μ_i, where n is an arbitrary integer larger than 1, and $i = 1, 2, \ldots, n$, such that for all n and for all $0 \leq \underline{\mu}_i \leq \overline{\mu}_i$, we have $\sum\limits_{i=1}^{n} f_{ni}(\underline{\mu}_1, \overline{\mu}_1, \ldots, \underline{\mu}_n, \overline{\mu}_n) = 1$.

Definition 14.7 We say that an interval-based portfolio allocation scheme is *symmetric* if for each n, for each $\underline{\mu}_1, \overline{\mu}_1, \ldots, \underline{\mu}_n, \overline{\mu}_n$, for each $i \leq n$, and for each permutation $\pi : \{1, \ldots, n\} \to \{1, \ldots, n\}$, we have

$$f_{ni}(\underline{\mu}_1, \overline{\mu}_1 \ldots, \underline{\mu}_n, \overline{\mu}_n) = f_{n,\pi(i)}(\underline{\mu}_{\pi(1)}, \overline{\mu}_{\pi(1)}, \ldots, \underline{\mu}_{\pi(n)}, \overline{\mu}_{\pi(n)}).$$

Definition 14.8 We say that an interval-based portfolio allocation scheme is *consistent* if for every $n > 2$ and for all $i \neq j$, we have

$$f_{ni}(\underline{\mu}_1, \overline{\mu}_1, \ldots, \underline{\mu}_n, \overline{\mu}_n) =$$

$$f_{21}(\underline{\mu}_i, \overline{\mu}_i, \underline{\mu}_j, \overline{\mu}_j) \cdot (f_{ni}(\underline{\mu}_1, \overline{\mu}_1, \ldots, \underline{\mu}_n, \overline{\mu}_n) + f_{nj}(\underline{\mu}_1, \overline{\mu}_1, \ldots, \underline{\mu}_n, \overline{\mu}_n)).$$

Proposition 14.4 *An interval-based portfolio allocation scheme is symmetric and consistent if and only if there exists a function $f(\underline{\mu}, \overline{\mu}) \geq 0$ for which*

$$f_{ni}(\underline{\mu}_1, \overline{\mu}_1, \ldots, \underline{\mu}_n, \overline{\mu}_n) = \frac{f(\underline{\mu}_i, \overline{\mu}_i)}{\sum\limits_{j=1}^{n} f(\underline{\mu}_j, \overline{\mu}_j)}.$$

Proof is similar to the proof of Proposition 14.1.

Definition 14.9 We say that an interval-based portfolio allocation scheme is *monotonic* if for each n and each $\underline{\mu}_i$ and $\overline{\mu}_i$, if $\underline{\mu}_i \geq \underline{\mu}_j$ and $\overline{\mu}_i \geq \overline{\mu}_j$, then

$$f_{ni}(\underline{\mu}_1, \overline{\mu}_1, \ldots, \underline{\mu}_n, \overline{\mu}_n) \geq f_{nj}(\underline{\mu}_1, \overline{\mu}_1, \ldots, \underline{\mu}_n, \overline{\mu}_n).$$

One can easily check that a symmetric and consistent portfolio allocation scheme is monotonic if and only if the corresponding function $f(\underline{\mu}, \overline{\mu})$ is non-decreasing in both variables.

Additivity. Let us assume that in year 1, we have instruments with bounds $\underline{\mu}_i$ and $\overline{\mu}_i$, and in year 2, we have a different set of instruments, with bounds $\underline{\mu}'_j$ and $\overline{\mu}'_j$. Then, we can view this situation in two different ways:

- we can view it as two different portfolio allocations, with allocations w_i in the first year and independently, allocations w'_j in the second year; since these two years are treated independently, the portion of money that goes into the i-th instrument in the first year and in the j-th instrument in the second year can be simply computed as a product $w_i \cdot w'_j$ of the corresponding portions;
- alternatively, we can consider portfolio allocation as a 2-year problem, with $n \cdot m$ possible options, so that for each option (i, j), the expected return is the sum $\mu_i + \mu'_j$ of the corresponding expected returns; since μ_i is in the interval $[\underline{\mu}_i, \overline{\mu}_i]$ and μ'_j is in the interval $[\underline{\mu}'_j, \overline{\mu}'_j]$, the sum $\mu_i + \mu'_j$ can take all the values from $\underline{\mu}_i + \underline{\mu}'_i$ to $\overline{\mu}_i + \overline{\mu}'_j$.

It is reasonable to require that the resulting portfolio allocation not depend on how exactly we represent this situation.

Definition 14.10 An interval-based portfolio allocation scheme is called *additive* if for every n and m, for all values $\underline{\mu}_i$, $\overline{\mu}_i$, $\underline{\mu}'_i$, and $\overline{\mu}'_i$, and for every i and j, we have

$$f_{n \cdot m, i, j}(\underline{\mu}_1 + \underline{\mu}'_1, \overline{\mu}_1 + \overline{\mu}'_1, \underline{\mu}_1 + \underline{\mu}'_2, \overline{\mu}_1 + \overline{\mu}'_2, \ldots, \underline{\mu}_n + \underline{\mu}'_m, \overline{\mu}_n + \overline{\mu}'_m) =$$

$$f_{ni}(\underline{\mu}_1, \overline{\mu}_1, \ldots, \underline{\mu}_n, \overline{\mu}_n) \cdot f_{mj}(\underline{\mu}'_1, \overline{\mu}'_1, \ldots, \underline{\mu}'_n, \overline{\mu}'_n).$$

Proposition 14.5 *A symmetric and consistent interval-based portfolio allocation scheme is additive if and only if the corresponding function $f(\underline{u}, \overline{u})$ has the form*

$$f(\underline{u}, \overline{u}) = \exp(\underline{\beta} \cdot \underline{u} + \overline{\beta} \cdot \overline{u})$$

for some $\underline{\beta} \geq 0$ and $\overline{\beta} \geq 0$.

Proof In terms of the function $f(\underline{u}, \overline{u})$, additivity takes the form

$$f(\underline{u} + \underline{u}', \overline{u} + \overline{u}') = C \cdot f(\underline{u}, \overline{u}) \cdot f(\underline{u}', \overline{u}').$$

For $F \stackrel{\text{def}}{=} \ln(f)$, this equation has the form

$$F(\underline{u} + \underline{u}', \overline{u} + \overline{u}') = c + F(\underline{u}, \overline{u}) + F(\underline{u}', \overline{u}'),$$

where $c \stackrel{\text{def}}{=} \ln(C)$. For $G \stackrel{\text{def}}{=} F + c$, we have

$$G(\underline{u} + \underline{u}', \overline{u} + \overline{u}') = G(\underline{u}, \overline{u}) + G(\underline{u}', \overline{u}').$$

According to [10], the only monotonic solution to this equation is a linear function. Thus, the function $f = \exp(F) = \exp(G - c) = \exp(-c) \cdot \exp(G)$ has the desired form. The proposition is proven.

Relation to Hurwicz approach to decision making under interval uncertainty. The above formula has the form $\exp(\beta \cdot (\alpha_H \cdot \overline{u} + (1 - \alpha_H) \cdot \underline{u}))$, where $\beta \stackrel{\text{def}}{=} \underline{\beta} + \overline{\beta}$ and $\alpha_H \stackrel{\text{def}}{=} \overline{\beta}/\beta$.

Thus, it is equivalent to using the non-interval formula with

$$u = \alpha_H \cdot \overline{u} + (1 - \alpha_H) \cdot \underline{u}.$$

This is exactly the utility equivalent to an interval proposed by a Nobelist Leo Hurwicz; see, e.g., [11–13].

Relation to maximum entropy. This formula corresponds to maximizing entropy under the constraint that the expected value of the Hurwicz combination $u = \alpha_H \cdot \overline{u} + (1 - \alpha_H) \cdot \underline{u}$ takes a given value.

References

1. L. Bokati, V. Kreinovich, Maximum entropy approach to portfolio optimization: economic justification of an intuitive diversity idea. Asian J. Econ. Bank. **3**(2), 17–28 (2019)
2. H.M. Markowitz, Portfolio selection. J. Financ. **7**(1), 77–91 (1952)
3. E.T. Jaynes, G.L. Bretthorst, *Probability Theory: The Logic of Science* (Cambridge University Press, Cambridge, UK, 2003)
4. M. Abbassi, M. Ashrafi, E. Tashnizi, Selecting balanced portfolios of R&D projects with interdependencies: a cross-entropy based methodology. Technovation **34**, 54–63 (2014)
5. A. Bera, S. Park, Optimal portfolio diversification using the maximum entropy principle. Econ. Rev. **27**(2–4), 484–512 (2008)
6. M. Sheraz, S. Dedu, V. Preda, Entropy measures for assessing volatile markets. Procedia Econ. Financ. **22**, 655–662 (2015)
7. J.R. Yu, W.Y. Lee, W.J.P. Chiou, Diversified portfolios with different entropy measures. Appl. Math. Comput. **241**, 47–63 (2014)
8. R. Zhou, R. Cai, G. Tong, Applications of entropy in finance: a review. Entropy **15**, 4909–4931 (2013)
9. R. Zhou, Y. Zhan, R. Cai, G. Tong, A mean-variance hybrid-entropy model for portfolio selection with fuzzy returns. Entropy **17**, 3319–3331 (2015)

10. J. Aczél, J. Dhombres, *Functional Equations in Several Variables*, (Cambridge University Press, 2008)
11. L. Hurwicz, *Optimality Criteria for Decision Making Under Ignorance*, vol. 370, (Cowles Commission Discussion Paper, Statistics, 1951)
12. V. Kreinovich, Decision making under interval uncertainty (and beyond), in *Human-Centric Decision-Making Models for Social Sciences*. ed. by P. Guo, W. Pedrycz (Springer Verlag, 2014), pp.163–193
13. R.D. Luce, R. Raiffa, *Games and Decisions: Introduction and Critical Survey* (Dover, New York, 1989)

Chapter 15
People Make Decisions Using Heuristics. II

In this chapter, we provide a justification for yet another heuristic that people use when making economic decision: the so-called anchoring effect. According to the traditional economics, the price that a person is willing to pay for an item should be uniquely determined by the value that this person will get from this item, it should not depend, e.g., on the asking price proposed by the seller. In reality, the price that a person is willing to pay *does* depend on the asking price; this is known as the anchoring effect. In this chapter, we provide a natural justification for the empirical formula that describes this effect.

Comment. Results from this chapter first appeared in [1].

15.1 Formulation of the Problem

What is anchoring effect? Traditional economics assumes that people know the exact value of each possible item, and this value determines the price that they are willing to pay for this item.

The reality is more complicated. In many practical situations, people are uncertain about the value of an item—and thus, uncertain about the price they are willing to pay for this item. This happens, e.g., when hunting for a house.

Interestingly, in many such situations, the price that the customer is willing to pay is affected by the asking price:

- if the asking price is higher, the customer is willing to pay a higher price, but
- if the asking price is lower, the price that the customer is willing to pay is also lower.

This phenomenon is known as the *anchoring effect*: just like a stationary ship may move a little bit, but cannot move too far away from its anchor, similarly, a customer

L. Bokati and V. Kreinovich, *Decision Making Under Uncertainty, with a Special Emphasis on Geosciences and Education*, Studies in Systems, Decision and Control 218,
https://doi.org/10.1007/978-3-031-26086-5_15

stays closer to the asking price—which thus acts as a kind of an anchor; see, e.g., [2], Chap. 11, and references therein.

Comment. The anchoring effect may sound somewhat irrational, but it makes some sense:

- If the owner lists his/her house at an unexpectedly high price, then maybe there are some positive features of the house of which the customer is not aware. After all, the owner does want to sell his/her house, so he/she would not just list an outrageously high price without any reason.
- Similarly, if the owner lists his/her house at an unexpectedly low price, then maybe there are some drawbacks of the house or of its location of which the customer is not aware. After all, the owner does want to get his/her money back when selling his/her house, so he/she would not just list an outrageously low price without any reason.

A formula that describes the anchoring effect. Let p_0 be the price that the customer would suggest in the absence of an anchor. Of course, if the asking price a_0 is the same value $a = p_0$, there is no reason for the customer to change the price p that he/she is willing to pay for this item, i.e., this price should still be equal to p_0.

It turns out that each anchoring situation can be described by a coefficient $\alpha \in [0, 1]$ which is called an *anchoring index*. The idea is that if we consider two different asking prices $a' \neq a''$, then the difference $p' - p''$ between the resulting customer's prices should be equal to $\alpha \cdot (a' - a'')$.

This idea—in combination with the fact that $p = p_0$ when $a = p_0$—enables us to come up with the formula describing the anchoring effect. Indeed, for anchor a, thre difference $p - p_0$ between:

- the price p corresponding to the asking price a and
- the price p_0 corresponding to the asking price p_0

should be equal to $\alpha \cdot (a - p_0)$. Since $p - p_0 = \alpha \cdot (a - p_0)$, we thus have $p = p_0 + \alpha \cdot (a - p_0)$, i.e., equivalently,

$$p = (1 - \alpha) \cdot p_0 + \alpha \cdot a. \tag{15.1}$$

First natural question: how can we explain this empirical formula?

What are the values of the anchoring index. It turns out that in different situations, we observe different values of the anchoring index.

When people are not sure about their original opinion, the anchoring index is usually close to 0.5:

- For a regular person buying a house, this index is equal to $0.48 \approx 0.5$; see, e.g., [2, 3].
- For people living in a polluted city, when asked what living costs they would accept to move to an environmentally clean area, the anchoring index was also close to 0.5; see, e.g., [2].

For other situations, when a decision maker in more confident in his/her original opinion, we can get indices between 0.25 and 0.5:

- For a real estate agent buying a house, this index is equal to 0.41; see, e.g., [2, 3].
- For a somewhat similar situation of charity donations, this index is equal to 0.30; see, e.g., [2, 4].

Second natural question: how can we explain these values?

What we do in this chapter. In this chapter, we try our best to answer both questions. Specifically:

- we provide a formal explanation for the formula (15.1), and
- we provide a somewhat less formal explanation for the empirically observed values of the anchoring index.

To make our explanations more convincing, we have tried to make the corresponding mathematics as simple as possible.

15.2 Formal Explanation of the Anchoring Formula

What we want. We want to have a function that, given two numbers:

- the price p_0 that the customer is willing to pay in a situation in which the seller has not yet proposed any asking price, and
- the actual asking price a,

produces the price $p(p_0, a)$ that the customer is willing to pay for this item after receiving the asking price a.

First natural property. As we have mentioned, if $a = p_0$, then we should have $p(p_0, a) = p(p_0, p_0) = p_0$.

Second natural property. Small changes in p_0 and a should not lead to drastic changes in the resulting price. In mathematical terms, this means that the function $p(p_0, a)$ should be continuous.

Third natural property. Intuitively, the change from p_0 to p should be in the direction to the anchor, i.e.:

- if $a < p_0$, we should have $p(p_0, a) \le p_0$, and
- if $p_0 < a$, we should have $p_0 \le p(p_0, a)$.

Fourth natural property. Also, intuitively, when the changed value $p(p_0, a)$ moves in the direction of the asking price a, it should not exceed a, i.e.:

- if $a < p_0$, we should have $a \le p(p_0, a)$, and
- if $p_0 < a$, we should have $p(p_0, a) \le a$.

Comment. The first three property can be summarized by saying that for all p_0 and a, the price $p(p_0, a)$ should always be in between the original price p_0 and the asking price a.

Fourth natural property: additivity. Suppose that we have two different situations—e.g., a customer is buying two houses, a house to live in and a smaller country house for vacationing. Suppose that:

- for the first item, the original price was p_0' and the asking price is a', and
- for the second item, the original price was p_0'' and the asking price is a''.

Then, the price of the first item is $p(p_0', a')$, the price of the second item is $p(p_0'', a'')$, thus the overall price of both items is

$$p(p_0', a') + p(p_0'', a''). \tag{15.2}$$

Alternatively, instead of considering the two items separately, we can view them as a single combined item, with the original price $p_0' + p_0''$ and the asking price $a' + a''$. From this viewpoint, the resulting overall price of both items is

$$p(p_0' + p_0'', a' + a''). \tag{15.3}$$

Since (15.2) and (15.3) correspond to the exact same situation, it is reasonable to require that these two overall prices should coincide, i.e., that we should have

$$p(p_0', a') + p(p_0'', a'') = p(p_0' + p_0'', a' + a''). \tag{15.4}$$

Now, we are ready to formulate and prove our main result.

Definition 15.1 A continuous function $p : \mathbb{R}_0^+ \times \mathbb{R}_0^+ \to \mathbb{R}_0^+$ that transforms two non-negative numbers p_0 and a into a non-negative number $p(p_0, a)$ is called an *anchoring function* if it satisfies the following two properties:

- for all p_0 and a, the value $p(p_0, a)$ should always be in between p_0 and a, and
- for all possible values p_0', p_0'', a', and a'', we should have

$$p(p_0', a') + p(p_0'', a'') = p(p_0' + p_0'', a' + a'').$$

Proposition 15.1 *A function $p(p_0, a)$ is an anchoring function if and only if it has the form*

$$p(p_0, a) = (1 - \alpha) \cdot p_0 + \alpha \cdot a$$

for some $\alpha \in [0, 1]$.

Comment. This proposition justifies the empirical expression (15.1) for the anchoring effect.

Proof It is easy to see that every function of the type (15.1) satisfies both conditions of Definition 15.1 and is, thus, an anchoring function. So, to complete the proof, it is sufficient to prove that every anchoring function—i.e., every function that satisfies both conditions from Definition 15.1—indeed has the form (15.1). □

Indeed, let us assume that the function $p(p_0, a)$ satisfies both conditions. Then, due to additivity, for each p_0 and a, we have

$$p(p_0, a) = p(p_0, 0) + p(0, a). \tag{15.5}$$

Thus, to find the desired function of two variables, it is sufficient to consider two functions of one variable: $p_1(p_0) \stackrel{\text{def}}{=} p(p_0, 0)$ and $p_2(a) \stackrel{\text{def}}{=} p(0, a)$.

Due the same additivity property, each of these functions is itself additive:

$$p(p_0' + p_0'', 0) = p(p_0', 0) + p(p_0'', 0)$$

and

$$p(0, a' + a'') = p(0, a') + p(0, a'').$$

In other word, both functions $p_1(x)$ and $p_2(x)$ are additive in the sense that for each of them, we always have $p_i(x' + x'') = p_i(x') + p_i(x'')$.

Since the function $p(p_0, a)$ is continuous, both functions $p_i(x)$ are continuous as well. Let us show that every continuous additive function is linear, i.e., has the form $p_i(x) = c_i \cdot x$ for some c_i.

Indeed, let us denote $c_i \stackrel{\text{def}}{=} p_i(1)$. Due to additivity, since

$$\frac{1}{n} + \cdots + \frac{1}{n} \ (n \text{ times}) = 1,$$

we have

$$p_i\left(\frac{1}{n}\right) + \cdots + p_i\left(\frac{1}{n}\right) \ (n \text{ times}) = p_i(1) = c_i,$$

i.e.,

$$n \cdot p_i\left(\frac{1}{n}\right) = c_i$$

and thus,

$$p_i\left(\frac{1}{n}\right) = c_i \cdot \frac{1}{n}.$$

Similar, due to additivity, since for every m and n, we have

$$\frac{1}{n} + \cdots + \frac{1}{n} \ (m \text{ times}) = \frac{m}{n},$$

we have

$$p_i\left(\frac{1}{n}\right)+\cdots+p_i\left(\frac{1}{n}\right)\ (m \text{ times}) = p_i\left(\frac{m}{n}\right).$$

The left-hand side of this formula is equal to

$$m\cdot p_i\left(\frac{1}{n}\right)=m\cdot\left(c_i\cdot\frac{1}{n}\right)=c_i\cdot\frac{m}{n}.$$

Thus, for every m and n, we have

$$p_i\left(\frac{m}{n}\right)=c_i\cdot\frac{m}{n}.$$

The property $p_i(x)=c_i\cdot x$ therefore holds for every rational number, and since each real number x can be viewed as a limit of its more and more accurate rational approximations x_n ($x=\lim x_n$), and the function $p_i(x)$ is continuous, we thus conclude, in the limit, that $p_i(x)=c_i\cdot x$ for all non-negative numbers x.

Thus, $p(p_0,0)=p_1(p_0)=c_1\cdot p_0$, $p(0,a)=p_2(a)=c_2\cdot a$, and the formula (15.5) takes the form

$$p(p_0,a)=c_1\cdot p_0+c_2\cdot a. \tag{15.6}$$

For $p_0=a$, the requirement that $p(p_0,a)$ is between p_0 and a implies that $p(p_0,a)=p_0$. For $p_0=a$, the formula (15.6) means that $c_1\cdot p_0+c_2\cdot p_0=p_0$, thus that $c_1+c_2=1$ and $c_1=1-c_2$. So, we get the desired formula (1) with $c_2=\alpha$.

To complete the proof, we need to show that $0\le\alpha\le 1$. Indeed, for $p_0=0$ and $a=1$, the value $p(0,1)$ must be between 0 and 1. Due to the formula (15.1), this value is equal to $(1-c_2)\cdot 0+c_2\cdot 1=c_2$. Thus, $c_2\in[0,1]$.

The proposition is proven.

15.3 Explaining the Numerical Values of the Anchoring Index

First case. Let us first consider the case when the decision maker is not sure which is more important: his/her a priori guess—as reflected by the original value p_0—or the additional information as described by the asking price a. In this case, in principle, the value α can take any value from the interval [0, 1].

To make a decision, we need to select one value α_0 from this interval. Let us consider the discrete approximation with accuracy $\frac{1}{N}$ for some large N. In this approximation, we only need to consider values

$$0,\frac{1}{N},\frac{2}{N},\ldots,\frac{N-1}{N},1,$$

for some large N. If we list all possible values, we get a tuple

$$\left(0, \frac{1}{N}, \frac{2}{N}, \ldots, \frac{N-1}{N}, 1\right).$$

We want to select a single tuple α_0, i.e., in other words, we want to replace the original tuple with a tuple $(\alpha_0, \ldots, \alpha_0)$. It is reasonable to select the value α_0 for which the replacing tuple is the closest to the original tuple, i.e., for which the distance

$$\sqrt{(\alpha_0 - 0)^2 + \left(\alpha_0 - \frac{1}{N}\right)^2 + \left(\alpha_0 - \frac{2}{N}\right)^2 + \cdots + \left(\alpha_0 - \frac{N-1}{N}\right)^2 + (\alpha_0 - 1)^2}$$

attains its smallest possible value. Minimizing the distance is equivalent to minimizing its square

$$(\alpha_0 - 0)^2 + \left(\alpha_0 - \frac{1}{N}\right)^2 + \left(\alpha_0 - \frac{2}{N}\right)^2 + \cdots + \left(\alpha_0 - \frac{N-1}{N}\right)^2 + (\alpha_0 - 1)^2.$$

Differentiating this expression with respect to α_0 and equating the derivative to 0, we conclude that

$$2\,(\alpha_0 - 0) + 2\left(\alpha_0 - \frac{1}{N}\right) + 2\left(\alpha_0 - \frac{2}{N}\right) + \cdots + 2\left(\alpha_0 - \frac{N-1}{N}\right) + 2\,(\alpha_0 - 1) = 0.$$

If we divide both sides by 2 and move the terms not containing α_0 to the right-hand side, we conclude that

$$(N+1) \cdot \alpha_0 = 0 + \frac{1}{N} + \frac{2}{N} + \cdots + \frac{N-1}{N} + 1,$$

i.e., that

$$(N+1) \cdot \alpha_0 = \frac{1 + 2 + \cdots + (N-1) + N}{N},$$

thus

$$\alpha_0 = \frac{1 + 2 + \cdots + (N-1) + N}{N \cdot (N+1)}.$$

It is known that $1 + 2 + \cdots + N = \dfrac{N \cdot (N+1)}{2}$, thus

$$\alpha_0 = 0.5.$$

This is exactly the value used when the decision maker is not confident in his/her original estimate.

Second case. What if the decision maker has more confidence in his/her original estimate than in the anchor? In this case, the weight $1 - \alpha$ corresponding to the original estimate must be larger than the weight α corresponding to the anchor. The inequality $1 - \alpha > \alpha$ means that $\alpha < 0.5$.

Similarly to the above case, we can consider all possible values between 0 and 0.5, and select a single value α_0 which is, on average, the closest to all these values. Similar to above calculations, we can conclude that the best value is

$$\alpha = 0.25.$$

Correspondingly, intermediate cases when the decision maker's confidence in his original opinion is somewhat larger, can be described by values α between the two above values 0.5 and 0.25. This explains why these intermediate values occur in such situations.

References

1. L. Bokati, V. Kreinovich, C. Van Le, How to explain the anchoring formula in behavioral economics, in *Prediction and Causality in Econometrics and Related Topics*, ed. by N.N. Thach, D.T. Ha, N.D. Trung, V. Kreinovich (Springer, Cham, 2022), pp. 28–34
2. D. Kahneman, *Thinking, Fast and Slow* (Farrar, Straus, and Giroux, New York, 2011)
3. G.B. Northcraft, M.A. Neale, Experts, amateurs, and real estate: an achoring-and-adjustment perspective on property pricing decisions. Organizat. Behav. Human Decis. Process. **39**, 84–97 (1987)
4. K.E. Jacowitz, D. Kahneman, Measures of anchoring in estimation tasks. Person. Soc. Psychol. Bull. **21**, 1161–1166 (1995)

Part III
Applications to Geosciences

After the general description of human decision making, in Part III we focus on our main application area: geosciences. In geosciences, like in many other application areas, we encounter two types of situations.

In some cases, we have a relatively small number of observations--only sufficiently many to estimate the values of a few parameters of the model. In such cases, it is desirable to come up with the most adequate few-parametric model.We analyze the corresponding problem of select an optimal model on two examples:

- of spatial dependence (Chap. 16) and
- of temporal dependence (Chap. 17).

As an example of a temporal dependence problem, we consider one of the most challenging and the most important geophysical problems: the problem of earthquake prediction. Specifically, we analyze the problem of selecting the most adequate probabilistic distribution of between-earthquakes time intervals.

In other cases, we already have many observations covering many locations and many moments of time. In such cases, we can look for the best ways to extend this knowledge:

- to other spatial locations (Chap. 18) and
- to future moments of time (Chap. 19).

As an example of extending knowledge to future moments of time-i.e., prediction-we deal with one of the least studied seismic phenomena: earthquakes triggering other earthquakes.

Chapter 16
Few-Parametric Spatial Models and How They Explain Bhutan Landscape Anomaly

Now that we have finished our general analysis of human decision making, let us move to our main application area: geosciences. In geosciences, like in many other application areas, we encounter two types of situations. In some cases, we have a small number of observations; in this case, it is important to extract as much information from these observations as possible. In other cases, we have a large number of observations—in such cases, we need to be able to process all this data in reasonable time.

Let us start with the cases when we have a relatively small number of observations. In such cases, there are only sufficiently many to estimate the values of a few parameters of the model. In such cases, it is desirable to come up with the most adequate few-parametric model. We analyze the corresponding problem of select an optimal model on two examples:

- of spatial dependence (in this chapter) and
- of temporal dependence (in the next chapter).

Let us explain a specific problem related to spatial dependence—and how it is related to economic decision making. Economies of countries located in seismic zones are strongly effected by this seismicity. If we underestimate the seismic activity, then a reasonably routine earthquake can severely damage the existing structures and thus, lead to huge economic losses. On the other hand, if we overestimate the seismic activity, we waste a lot of resources on unnecessarily fortifying all the buildings—and this too harms the economies. From this viewpoint, it is desirable to have estimations of regional seismic activities which are as accurate as possible. Current predictions are mostly based on the standard geophysical understanding of earthquakes as being largely caused by the movement of tectonic plates and terranes. This understanding works in most areas, but in Bhutan area of the Himalayas region, there seems to be a landscape anomaly. As a result, for this region, we have less confidence in the accuracy of seismic predictions based on the standard understanding and thus, have to use higher seismic thresholds in construction. In this chapter, we find the optimal description of landscape-describing elevation profiles, and we use

L. Bokati and V. Kreinovich, *Decision Making Under Uncertainty, with a Special Emphasis on Geosciences and Education*, Studies in Systems, Decision and Control 218,
https://doi.org/10.1007/978-3-031-26086-5_16

this description to show that the seeming anomaly is actually in perfect agreement with the standard understanding of the seismic activity. Our conclusion is that it is safe to apply, in this region, estimates based on the standard understanding and thus, avoid unnecessary expenses caused by an increased threshold.

Comment. Results from this chapter first appeared in [1].

16.1 Formulation of the Problem

Seismicity affects economy. In highly seismic areas like the Himalayas, economy is affected by our knowledge of possible seismicity.

Protection against possible earthquakes is very costly. If we have only a vague idea about possible seismic events—i.e., if we can potentially expect high-energy earthquakes at all possible locations—then, every time we build a house or a factory, we need to spend a lot of money on making it protected against such events—with little money left for any other development project.

On the other hand, if we can reasonably accurately localize potential hazards, then we can concentrate our building efforts mostly in safer zones. This will require less investment in earthquake protection and thus, leave more money for other development projects.

Thus, the economy of a highly seismic zone is directly affected by our understanding of the corresponding seismic processes.

Bhutan landscape anomaly. In general, modern geophysics has a reasonably good understanding of seismic processes and seismic zones. Specifically, the current understanding is that seismicity is usually caused by mutual movement of tectonic plates and their parts (terranes), and it is mostly concentrated on the borderline between two or more such plates or terranes. In general, while we still cannot predict the exact timing of earthquakes, geoscientists can reasonably well predict the size of a future earthquake based on the corresponding geophysical models.

Researchers and practitioners are reasonably confident in these predictions—at least for locations whose geophysics is well understood by the traditional geophysical models.

However, there are locations where observed phenomena are different from what we usually expect. In such cases, there are reasonable doubts in seismicity estimates produced by the traditional techniques—and thus, it is reasonable to be cautious and use higher strengths of potential earthquakes when building in these locations, which invokes significant additional expenses. For such domains, it is therefore desirable to come up with a better understanding of the observed geophysical phenomena—thus hopefully allowing us to make more accurate predictions and hence, save money (which is now wasted on possibly too-heavy earthquake protection) for other important activities.

One such areas in the vicinity of the Himalayan country of Bhutan, where the landscape profile is drastically different from the profiles of other Himalayan areas such as areas of Nepal. In general, a landscape can be described in numerical terms if we take a line orthogonal to the prevailing rivers (which are usually the lowest points on the landscape) and plot the elevation as a function of the distance from the corresponding river. The shape of the landscape (elevation) profile in Bhutan is visually drastically different from the landscape profile in Nepal; see, e.g., [2]. Namely, in most of the Himalayas—and, in general, in the most of the world—the corresponding curve is first convex (corresponding to the river valley), and then becomes concave—which corresponds to the mountain peaks. In contrast, in Bhutan, the profile turns concave very fast, way before we reach the mountain peaks area.

As of now, there are no good well-accepted explanations for this phenomenon—which makes it an anomaly. To be more precise, we know that the geophysics of the Bhutan area is somewhat different: in Nepal (like in most areas in the world), the advancing tectonic plate in orthogonal to the border of the mountain range, while in Bhutan, the plate pushes the range at an angle. However, it is not clear how this can explain the above phenomenon. This leads us to the following questions.

Questions. The first question is: can we explain the Bhutan anomaly within the existing geophysical paradigm? If we can, this would mean that this anomaly is not an obstacle to applying this paradigm, and thus, that the estimates of future seismic activity obtained within this paradigm can be safely applied—without the need to make expensive extra precautions.

A related question is related to the fact that while we use convexity and concavity to describe elevation profiles, the only reason for using these two properties is because these are the basic properties that we learn in math. Is there any geophysical meaning in convexity vs. concavity?

What we do in this chapter. In this chapter, we provide answers to both questions: we explain why convexity and concavity are adequate ways to describe elevation profiles, and we explain how the at-an-angle pressure in the Bhutan area leads to the observed convex-followed-by-concave phenomenon.

To answer these questions, we first formulate the problem of adequately describing elevation profiles as an optimization problem. Then, we solve this problem, and use the solution to answer the above two questions.

16.2 What Is the Optimal Description of Elevation Profiles

How can we describe elevation profiles? An elevation profile results from the joint effect of many different physical processes, from movement of tectonic plates to erosion. These process are largely independent from each other: e.g., erosion works the same way whether we have the landscape on the sea level or the same landscape which the geological processes raised to some elevation. Because of this independence, the observed profile $f(x)$ can be reasonably well represented as the sum of profiles corresponding to different processes:

$$f(x) = f_1(x) + \cdots + f_n(x).$$

Different profile-changing processes may have different intensity. So, to describe the effect of the i-th process, instead of a fixed function $f_i(x)$, it is more appropriate to use the correspondingly re-scaled term $C_i \cdot f_i(x)$, where the coefficients C_i describe the intensity of the i-th process, so that

$$f(x) = C_1 \cdot f_1(x) + \cdots + C_n \cdot f_n(x).$$

Due to erosion, discontinuities in the elevation profiles are usually smoothed out, so we can safely assume that the corresponding functions $f_i(x)$ are smooth (differentiable).

For such families, the problem of selecting the optimal description was formulated and solved in the Appendix. The result was that the optimal approximating family is the family of polynomials.

16.3 Why Convexity and Concavity Are Important in Elevation Profiles: An Explanation Based on the Optimality Result

Discussion. The above result provides us, for different n, with families of approximations to the elevation profiles. Let us start with the simplest possible approximation.

For $n = 1$, we get the class of constant functions—no landscape at all. For $n = 2$, we get a class of linear functions—no mountains, no ravines, just a flat inclined surface. So, the only non-trivial description of a landscape starts with $n = 3$, i.e., with quadratic functions.

We want to provide a qualitative classification of all such possible elevation functions. It is reasonable to say that the two elevation functions are equivalent if they differ only by re-scaling and shift of x and y:

Definition 16.1 We say that two quadratic functions $f(x)$ and $g(x)$ are *equivalent* if for some values $\lambda_x > 0$, $\lambda_y > 0$, x_0, and y_0, we have

$$g(x) = \lambda_y \cdot f(\lambda_x \cdot x + x_0) + y_0$$

for all x.

Proposition 16.1 *Every non-linear quadratic function is equivalent either to x^2 or to $-x^2$.*

Discussion. Thus, in this approximation, we have, in effect, two shapes: the shape corresponding to x^2 (convex) and the shape corresponding to $-x^2$ (concave). This result explains why our visual classification into convex and concave shapes makes perfect sense.

Proof Every non-linear quadratic function $g(x)$ has the form

$$g(x) = a_0 + a_1 \cdot x + a_2 \cdot x^2,$$

for some $a_2 \neq 0$.

If $a_2 > 0$, then this function can be represented as

$$a_2 \cdot \left(x + \frac{a_1}{2a_2}\right)^2 + \left(a_0 - \frac{a_1^2}{4a_2}\right),$$

i.e., can be represented in the desired form, with $f(x) = x^2$, $\lambda_x = 1$, $\lambda_y = a_2$, $x_0 = \frac{a_1}{2a_2}$, and $y_0 = a_0 - \frac{a_1^2}{4a_2}$.

If $a_2 < 0$, then this function can be represented as

$$|a_2| \cdot \left(-\left(x + \frac{a_1}{2a_2}\right)^2\right) + \left(a_0 - \frac{a_1^2}{4a_2}\right),$$

i.e., can be represented in the desired form, with $f(x) = -x^2$, $\lambda_x = 1$, $\lambda_y = |a_2|$, $x_0 = \frac{a_1}{2a_2}$, and $y_0 = a_0 - \frac{a_1^2}{4a_2}$. □

The proposition is proven.

16.4 Bhutan Anomaly Explained

Discussion. In the previous text, we have shown that the optimal description of an elevation profiles is by polynomials of a fixed degree.

In the first approximation, a landscape profile can be described by a quadratic function. To get a more accurate description, let us also consider cubic terms, i.e., let us consider profiles of the type

$$f(x) = a_0 + a_1 \cdot x + a_2 \cdot x^2 + a_3 \cdot x^3. \tag{16.1}$$

As a starting point $x = 0$ for the elevation profile, it makes sense to select the lowest (or the highest) point. In both cases, according to calculus, the first derivative of the elevation profile is equal to 0 at this point: $f'(0) = 0$. Substituting the above expression for $f(x)$ into this formula, we conclude that $a_1 = 0$ and thus,

$$f(x) = a_0 + a_2 \cdot x^2 + a_3 \cdot x^3. \tag{16.2}$$

Let us analyze how this approximation works for the above two cases: the case of Nepal and the case of Bhutan.

Case of Nepal. In the case of Nepal, the forces compressing the upper plate are orthogonal to the line of contact. This means that in this case, the forces do not change if we change left to right and right to left.

Since the whole mountain range was created by this force, it is reasonable to conclude that the corresponding elevation profile is also invariant with respect to swapping left and right, i.e., with respect to the transformation $x \to -x$:

$$f(x) = f(-x). \tag{16.3}$$

Substituting the cubic expression (16.2) for the profile $f(x)$ into this formula, we conclude that $a_3 = 0$. Thus, in this case, the elevation profile is quadratic even in this next approximation—and is, therefore, either convex or concave.

Case of Bhutan. In the case of Bhutan, the force is applied at an angle. Here, there is no symmetry with respect to $x \to -x$, so, in general, we have $a_3 \neq 0$. Thus, the second derivative—that describes whether a function is locally convex (when this second derivative is positive) or locally concave (when the derivative is negative)—becomes a linear function $6a_3 \cdot x + 2a_2$, with $a_3 \neq 0$.

A non-constant linear function always changes signs—this explains why in the case of Bhutan, convexity follows by concavity.

16.5 Auxiliary Question: How to Best Locate an Inflection Point

Practical problem. Many geophysical ideas are applicable only to valley-type convex domains or only to mountain-type concave domains. So, to apply these ideas to a real-life landscape, it is necessary to divide the whole landscape into convex and concave zones. What is the best way to do it? In other words, what is the best way to locate an *inflection point*, i.e., the point at which local convexity changes to local concavity?

First idea: a straightforward least squares approach. The first natural idea—motivated by the above analysis—is to approximate the actual elevation profile by a cubic function (16.1). The corresponding coefficients c_0, c_1, c_2, and c_3 can be obtained, e.g., by applying the least squares method to the corresponding system of linear equations

$$y_i \approx c_0 + c_1 \cdot x_i + c_2 \cdot x_i^2 + c_3 \cdot x_i^3,$$

where x_i is the i-th location and y_i is the i-th elevation.

The least squares method minimizes the sum

$$\sum_i (y_i - (c_0 + c_1 \cdot x_i + c_2 \cdot x_i^2 + c_3 \cdot x_i^3))^2.$$

Differentiating this expression with respect to each of the unknowns c_j and equating all four derivatives to 0, we get an easy-to-solve system of four linear equations with four unknowns.

Once we find the characteristics, we then estimate the location of the inflection point as the value at which the second derivative is equal to 0, i.e., the value $x_{\text{infl}} = -\frac{c_2}{3c_3}$.

Second idea: a model-free least squares approach. Instead of restricting ourselves to a cubic approximation, we can consider general convex functions. For a function $f(x)$ defined by its values $y_1 = f(x_1)$, $y_2 = f(x_2)$, ..., on a equally spaced grid

$$x_1, x_2 = x_1 + \Delta x, x_3 = x_1 + 2\Delta x, \ldots, x_N,$$

convexity is equivalent to the sequence of inequalities

$$y_i \le \frac{y_{i-1} + y_{i+1}}{2}. \tag{16.4}$$

For each set of actual profile points $\widetilde{y}_i$, we can therefore find the closest convex profile by looking for the values y_i that minimize the mean square error (MSE)

$$\frac{1}{N} \cdot \sum_i (\widetilde{y}_i - y_i)^2$$

under the constraints (16.4). The minimized expression is a convex function of the unknowns y_i, and each constraint—and thus, their intersection—defines a convex set. Thus, we can find the corresponding minimum by using a known algorithm for convex optimization (= minimizing a convex function on a convex domain); see, e.g., [3–5].

By applying this algorithm to actually convex profiles, we can find the largest and thus, the corresponding MSE. Let us denote the largest of such values by M. Then, to find an inflection point, we can consider larger and larger fragments of the original series $f(x_1)$, $f(x_2)$, ..., until we reach a point at which the corresponding MSE exceeds M. This is the desired inflection point.

We can speed up this algorithm if instead of slowly increasing the size of the still-convex fragment, we use bisection. Specifically, we always keep two values $\underline{p}$ and $\overline{p}$ such that the fragment until $\underline{p}$ is convex (within accuracy M), while the fragment up to the point $\overline{p}$ is not convex within the given accuracy.

In the beginning, we first apply our criterion to the whole list of N values. If the result is M-close to convex, we consider the profile convex—no inflection point here. If the result is not M-convex, then we take $\underline{p} = 1$ and $\overline{p} = N$.

Once we have two values $\underline{p} < \overline{p}$, we then take a midpoint $m \stackrel{\text{def}}{=} \frac{\underline{p} + \overline{p}}{2}$. If the segment up to this midpoint is M-convex, then we replace $\underline{p}$ with m. If this segment is not M-convex, we replace $\overline{p}$ with m.

In both case, we get a new interval $[\underline{p}, \overline{p}]$ whose width decreased by a factor of two. We started with width N. Thus, in $\log_2(N)$ steps, this size decreases to $N/2^{\log_2(N)} = N/N = 1$, i.e., we get the exact location of the inflection point.

Comment. Other algorithms for detecting inflection points are described, e.g., in [6, 7].

References

1. T.N. Nguyen, L. Bokati, A. Velasco, V. Kreinovich, Bhutan landscape anomaly: possible effect on himalayan economy (in view of optimal description of elevation profiles). Thai J. Math., Spec. Issue Struct. Change Model. Optim. Econom. 57–69
2. B.A. Adams, K.X. Whipple, K.V. Hodges, A.M. Heimsath, In situ development of high-elevation, low-relief landscapes via duplex deformation in the Eastern Himalayan hinterland, Bhutan. J. Geophys. Res. Earth Surf. **121**, 294–319 (2016)
3. J. Nocedal, S. Wright, *Numerical Optimization* (Springer, New York, 2006)
4. P. Pardalos, *Complexity in Numerical Optimization* (World Scientific, Singapore, 1993)
5. R.T. Rockafeller, *Convex Analysis* (Princeton University Press, Princeton, 1997)
6. N.K. Kachouie, A. Schwartzman, Non-parametric estimation of a single inflection point in noisy observed signal. J. Electr. Electron. Syst. **2**(2), Paper 1000108 (2013)
7. D. Manocha, J.F. Canny, Detecting cusps and inflection points in curves. Comput. Aid. Geom. Design **9**(1), 1–24 (1992)

Chapter 17
Few-Parametric Temporal Models and How They Explain Gamma Distribution of Seismic Inter-Event Times

In this chapter, we consider the problem of selecting an optimal model of few-parametric temporal dependence. As a case study, we take a problem related to one of the most challenging—and the least successful so far—aspects of geosciences: namely, a problem related to earthquake prediction.

Specifically, it is known that the distribution of seismic inter-event times is well described by the Gamma distribution. Recently, this fact has been used to successfully predict major seismic events. In this chapter, we explain that the Gamma distribution of seismic inter-event times can be naturally derived from the first principles.

Comment. Results from this chapter first appeared in [1].

17.1 Formulation of the Problem

Gamma distribution of seismic inter-event times: empirical fact. Detailed analysis of the seismic inter-event times t—i.e., of times between the two consequent seismic events occurring in the same area—shows that these times are distributed according to the Gamma distribution, with probability density

$$\rho(t) = C \cdot t^{\gamma-1} \cdot \exp(\mu \cdot t), \tag{17.1}$$

for appropriate values γ, μ, and C; see, e.g., [2, 3].

Lately, there has been a renewed interest in this seemingly very technical result, since a recent chapter [4] has shown that the value of the parameter μ can be used to predict a major seismic event based on the preceding foreshocks. Specifically, it turns out that more than 70% of major seismic events in Southern California could be predicted some time in advance—with an average of about two weeks in advance.

L. Bokati and V. Kreinovich, *Decision Making Under Uncertainty, with a Special Emphasis on Geosciences and Education*, Studies in Systems, Decision and Control 218,
https://doi.org/10.1007/978-3-031-26086-5_17

Why gamma distribution? This interest raises a natural question: why the inter-event times follow gamma distribution? In this chapter, we provide a possible theoretical explanation for this empirical fact.

17.2 Our Explanation

Maximum entropy: general idea. In our explanation, we will use Laplace's Indeterminacy Principle, which is also known as the maximum entropy approach; see, e.g., [5]. The simplest case of this approach is when we have n alternatives, and we have no reasons to believe that one of them is more probable. In this case, a reasonable idea is to consider these alternatives to be equally probable, i.e., to assign, to each of these n alternatives, the same probability $p_1 = \cdots = p_n$. Since the probabilities should add to 1, i.e., $\sum\limits_{i=1}^{n} p_i = 1$, we thus get $p_i = 1/n$.

In this case, we did not introduce any new degree of certainty into the situation that was not there before—as would have happened, e.g., if we selected a higher probability for one of the alternatives. In other words, out of all possible probability distributions, i.e., out of all possible tuples $(p_1, \ldots, p_n)$ for which $\sum\limits_{i=1}^{n} p_i = 1$, we selected the only one with the largest possible uncertainty.

In general, it is known that uncertainty can be described by the *entropy* S, which in the case of finitely many alternatives has the form $S = -\sum\limits_{i=1}^{n} p_i \cdot \ln(p_i)$, and in the case of a continuous random variable with probability density $\rho(x)$, for which sum becomes an integral, a similar form

$$S = -\int \rho(x) \cdot \ln(\rho(x))\, dx. \tag{17.2}$$

Maximum entropy: examples. If the only information that we have about a probability distribution is that it is located somewhere on the given interval $[a, b]$, then the only constraint on the corresponding probability density function is that the overall probability over this interval is 1, i.e., that

$$\int_a^b \rho(x)\, dx = 1. \tag{17.3}$$

So, to apply the maximum entropy approach, we need to maximize the objective function (17.2) under the constraint (17.3). The usual way of solving such constraint optimization problem is to apply Lagrange multiplier method that reduces the original constraint optimization problem to an appropriate unconstrained optimization problem. In this case, this new problem means maximizing the expression

$$-\int \rho(x) \cdot \ln(\rho(x))\, dx + \lambda \cdot \left(\int_a^b \rho(x)\, dx - 1 \right), \tag{17.4}$$

where the parameter λ—known as *Lagrange multiplier*—needs to be determined from the condition that the solution to this optimization problem satisfies the constraint (17.3).

For this problem, the unknowns are the values $\rho(x)$ corresponding to different x. Differentiating the expression (17.4) with respect to $\rho(x)$ and equating the derivative to 0, we get the following equation:

$$-\ln(\rho(x)) - 1 + \lambda = 0.$$

(Strictly speaking, we need to use variational differentiation, since the unknown is a function.) The above equation implies that $\ln(\rho(x)) = \lambda - 1$, and thus, that $\rho(x) = \text{const}$. So, we get a uniform distribution—in full accordance with the original idea that, since we do not have any reasons to believe that some points on this interval are more probable than others, we consider all these points to be equally probable.

If, in addition to the range $[a, b]$, we also know the mean value

$$\int_a^b x \cdot \rho(x)\, dx = m \tag{17.5}$$

of the corresponding random variable, then we need to maximize the entropy (17.2) under two constraints (17.3) and (17.5). In this case, the Lagrange multiplier method leads to the unconstrained optimization problem of maximizing the following expression:

$$-\int \rho(x) \cdot \ln(\rho(x))\, dx + \lambda \cdot \left(\int_a^b \rho(x)\, dx - 1 \right) + \lambda_1 \cdot \left(\int_a^b x \cdot \rho(x)\, dx - m \right). \tag{17.6}$$

Differentiating this expression with respect to $\rho(x)$ and equating the derivative to 0, we get the following equation:

$$-\ln(\rho(x)) - 1 + \lambda + \lambda_1 \cdot x = 0,$$

hence $\ln(\rho(x)) = (\lambda - 1) + \lambda_1 \cdot x$ and so, we get a (restricted) Laplace distribution, with the probability density $\rho(x) = C \cdot \exp(\mu \cdot x)$, where we denoted $C \stackrel{\text{def}}{=} \exp(\lambda - 1)$ and $\mu \stackrel{\text{def}}{=} \lambda_1$.

Comment. It is worth mentioning that if, in addition to the mean value, we also know the second moment of the corresponding random variable, then similar arguments lead us to a conclusion that the corresponding distribution is Gaussian. This conclusion is in good accordance with the ubiquity of Gaussian distributions.

What are the reasonable quantities in our problem. We are interested in the probability distribution of the inter-event time t. Based on the observations, we can find the mean inter-event time, so it makes sense to assume that we know the mean value of this time.

Usual (astronomical) time versus internal time: general idea. This mean value is estimated if we use the usual (astronomical) time t, as measured, e.g., by rotation of the Earth around its axis and around the Sun. However, it is known that many processes also have their own "internal" time—based on the corresponding internal cycles. For example, we can measure the biological time of an animal (or a person) by such natural cyclic activities as breathing or heartbeat. Usually, breathing and heart rate are more or less constant, but, e.g., during sleep, they slow down—as most other biological processes slow down. On the other hand, in stressful situations, e.g., when the animal's life is in danger, all the biological processes speed up—including breathing and heart rate. To adequately describe how different biological characteristics change with time, it makes sense to consider not only how they change in astronomical time, but also how they change in the corresponding internal time—measured not by number of Earth's rotations around the sun, but rather in terms of number of heartbeats. An even more drastic slowdown occurs when an animal hibernates. In general, the system's internal time can be sometimes slower than astronomical time, and sometimes faster.

Usual (astronomical) time versus internal time: case of seismic events. In our problem, there is a similar phenomenon: usually, seismic events are reasonably rare. However, the observations indicate that the frequency with which foreshocks appear increases when we get closer to a major seismic event. In such situation, the corresponding seismic processes speed up, so we can say that the internal time speeds up. In general, an internal time is often a more adequate description of the system's changes than astronomical time. It is therefore reasonable to supplement the mean value of the inter-event time measured in astronomical time by the mean value of the inter-event time measured in the corresponding internal time.

How internal time depends on astronomical time: general idea. To describe this idea in precise terms, we need to know how this internal time τ depends on the astronomical time. As we have mentioned, the usual astronomical time is measured by natural cycles, i.e., by processes which are periodic in terms of the time t. So, to find the expression for internal time, we need to analyze what cycles naturally appear in the studied system—and then define internal time in terms of these cycles.

To describe the system's dynamics means to describe how the corresponding physical quantities $x(t)$ change with time t. In principle, in different physical situations, we can have different functions $x(t)$. In principle, to describe a general function, we need to have infinitely many parameters—e.g., we need to describe the values of this function at different moments of time. In practice, however, we can only have finitely many parameters. So, it is reasonable to consider finite-parametric families of functions. The simplest—and most natural—is to select some basic functions $e_1(t), \ldots, e_n(t)$, and to consider all possible linear combinations of these functions, i.e., all possible functions of the type

$$x(t) = C_1 \cdot e_1(t) + \cdots + C_n \cdot e_n(t), \tag{17.6}$$

where $C_1, \ldots, C_n$ are the corresponding parameters. This is indeed what is done in many situations: sometimes, we approximate the dynamics by polynomials—linear combinations of powers t^k, sometimes we use linear combinations of sinusoids, sometimes linear combinations of exponential functions, etc.

How internal time depends on astronomical time: case of seismic events. The quality of this approximation depends on how adequate the corresponding basis functions are for the corresponding physical process. Let us analyze which families are appropriate for our specific problem: analysis of foreshocks preceding a major seismic event. In this analysis, we can use the fact that, in general, to transform a physical quantity into a numerical value, we need to select a starting point and a measuring unit. If we select a different starting point and/or a different measuring unit (e.g., minutes instead of seconds), we will get different numerical values for the same quantity.

For the inter-event times, the starting point is fixed: it is 0, the case when the next seismic events follows immediately after the previous one. So, the only remaining change is the change of a measuring unit. If we replace the original time unit with a one which is r times smaller, then all numerical values are multiplied by r, i.e., instead of the original value t, we get a new value $t_{\text{new}} = r \cdot t$. For example, if we replace minutes by seconds, then the numerical values of all time intervals are multiplied by 60, so that, e.g., 2.5 min becomes $60 \cdot 2.5 = 150$ s.

Some seismic processes are faster, some are slower. This means that, in effect, they differ by this slower-to-faster or faster-to-slower transformations $t \to r \cdot t$. We would like to have a general description that would fit all these cases. In other words, we would like to make sure that the class (17.6) remains the same after this "re-scaling", i.e., that for each i and for each r, the re-scaled function $e_i(r \cdot t)$ belongs to the same class (17.6). In other words, we require that for each i and r, there exists values $C_{ij}(r)$ for which

$$e_i(r \cdot t) = C_{i1}(r) \cdot e_1(t) + \cdots + C_{in}(r) \cdot e_n(t). \tag{17.7}$$

Let us solve the resulting systems of equations. Seismic waves may be changing fast but, in general, they are still smooth. It is therefore reasonable to consider only smooth functions $e_i(t)$. If we pick n different values $t_1, \ldots, t_n$, then, for each r and for each i, we get a system of n linear equations for determining n unknowns $C_{i1}(r)$, ..., $C_{in}(r)$:

$$e_i(r \cdot t_1) = C_{i1}(r) \cdot e_1(t_1) + \cdots + C_{in}(r) \cdot e_n(t_1);$$

$$\ldots$$

$$e_i(r \cdot t_n) = C_{i1}(r) \cdot e_1(t_n) + \cdots + C_{in}(r) \cdot e_n(t_n).$$

Due to Cramer's rule, each component $C_{ij}(r)$ of the solution to this system of linear equations is a ratio of two determinants and is thus, a smooth function of the corresponding coefficients $e_i(r \cdot t_j)$ and $e_i(t_j)$. Since the function $e_i(t)$ is differentiable, we conclude that the functions $C_{ij}(r)$ are also differentiable.

Since all the functions $e_i(t)$ and $C_{ij}(r)$ are differentiable, we can differentiate both sides of the formula (17.7) with respect to r and get:

$$e_i'(r \cdot t) \cdot t = C_{i1}'(r) \cdot e_1(t) + \cdots + C_{in}'(r) \cdot e_n(t),$$

where for each function f, the expression f', as usual, denotes the derivative. In particular, for $r = 1$, we get

$$e_i'(t) \cdot t = c_{i1} \cdot e_1(t) + \cdots + c_{in} \cdot e_n(t),$$

where we denoted $c_{ij} \stackrel{\text{def}}{=} C_{ij}'(1)$. For the auxiliary variable $T = \ln(t)$ for which $t = \exp(T)$, we have $dT = \dfrac{dt}{t}$, hence $\dfrac{de_i(t)}{dt} \cdot t = \dfrac{de_i(t)}{dT}$. So, for the auxiliary functions $E_i(T) \stackrel{\text{def}}{=} e_i(\exp(T))$, we get

$$E_i'(T) = c_{i1} \cdot E_1(T) + \cdots + c_{in} \cdot E_n(T).$$

So, for the functions $E_i(T)$, we get a system of linear differential equations with constant coefficients. It is well known that a general solution to such a system is a linear combination of the expressions $T^k \cdot \exp(a \cdot T) \cdot \sin(\omega \cdot T + \varphi)$ for some natural number k and real numbers a, ω, and φ. Thus, each function $e_i(t) = E_i(\ln(t))$ is a linear combination of the expressions

$$(\ln(t))^k \cdot \exp(a \cdot \ln(t)) \cdot \sin(\omega \cdot \ln(t) + \varphi) = (\ln(t))^k \cdot t^a \cdot \sin(\omega \cdot \ln(t) + \varphi). \tag{17.8}$$

So, the general expression (17.6) is also a linear combination of such functions.

The periodic part of this expression is a sine or cosine function of $\ln(t)$, so we can conclude that for seismic processes, the internal time τ is proportional to the logarithm $\ln(t)$ of the astronomic time: $\tau = c \cdot \ln(t)$ for some constant c.

This explains the ubiquity of Gamma distributions. Indeed, suppose that we know the mean values m_t and m_τ of the astronomical time t and the mean value of the internal time $\tau = c \cdot \ln(t)$. This means that the corresponding probability density function $\rho(t)$, in addition to the usual constraint $\int \rho(t)\,dt$, also satisfies the constraints

$$\int t \cdot \rho(t)\,dt = m_t$$

and

$$\int c \cdot \ln(t) \cdot \rho(t)\,dt = m_\tau.$$

Out of all possible distributions satisfying these three inequalities, we want to select the one that maximizes entropy

$$-\int \rho(t) \cdot \ln(\rho(t))\, dt.$$

To solve the resulting constraint optimization problem, we can apply the Lagrange multiplier method and reduce it to the unconstrained optimization problem of maximizing the expression:

$$-\int \rho(t) \cdot \ln(\rho(t))\, dt + \lambda \cdot \left(\int \rho(t)\, dt - 1\right) +$$

$$\lambda_t \cdot \left(\int t \cdot \rho(t)\, dt - m_t\right) + \lambda_\tau \cdot \left(\int c \cdot \ln(t) \cdot \rho(t)\, dt - m_\tau\right),$$

for some values λ, λ_t, and λ_τ. Differentiating both sides with respect to each unknown $\rho(t)$, we conclude that

$$-\ln(\rho(t)) - 1 + \lambda + \lambda_t \cdot t + \lambda_\tau \cdot c \cdot \ln(t) = 0,$$

i.e., that

$$\ln(\rho(t)) = (\lambda - 1) + \lambda \cdot t + (\lambda_\tau \cdot c) \cdot \ln(t).$$

Exponentiating both sides, we get the desired Gamma distribution form (17.1).

$$\rho(t) = C \cdot \tau^{\gamma-1} \cdot \exp(\mu \cdot t),$$

with $C = \exp(\lambda - 1)$, $\gamma = \lambda_\tau \cdot c + 1$, and $\mu = \lambda_t$. Thus, we have indeed explained the ubiquity of the Gamma distribution.

References

1. L. Bokati, A. Velasco, V. Kreinovich, Why Gamma distribution of seismic inter-event times: a theoretical explanation, in *How Uncertainty-Related Ideas Can Provide Theoretical Explanation for Empirical Dependencies*. ed. by M. Ceberio, V. Kreinovich (Springer, Cham, 2021), pp.43–50
2. A. Corral, Long-term clustering, scaling, and universality in the temporal occurrence of earthquakes. Phys. Rev. Lett. **92**, Paper 108501 (2004)
3. S. Heinzl, F. Scherbaum, C. Beauval, Estimating background activitiy based on interevent-time distribution. Bull. Seismol. Soc. Am. **96**(1), 313–320 (2006)
4. D.T. Trugman, Z.E. Ross, Pervasive foreshock activity across Southern California. Geophys. Res. Lett. **46** (2019). https://doi.org/10.1029/2019GL083725
5. E.T. Jaynes, G.L. Bretthorst, *Probability Theory: The Logic of Science* (Cambridge University Press, Cambridge, 2003)

Chapter 18
Scale-Invariance Explains the Empirical Success of Inverse Distance Weighting and of Dual Inverse Distance Weighting in Geosciences

Let us now consider the cases when we have a large number of observations and measurement results. Since we have many observations that cover many possible situations, a natural idea is to use these observations to predict what will happen in other cases. We can do it on a purely mathematical level–by simply interpolating the known observations, or we can also take into account the corresponding physics and related causality relation. In this chapter, we analyze the interpolation scenarios; in the following two chapters, we analyze how to take causality into account.

In general, once we measure the values of a physical quantity at certain spatial locations, we need to interpolate these values to estimate the value of this quantity at other locations x. In geosciences, one of the most widely used interpolation techniques is inverse distance weighting, when we combine the available measurement results with the weights inverse proportional to some power of the distance from x to the measurement location. This empirical formula works well when measurement locations are uniformly distributed, but it leads to biased estimates otherwise. To decrease this bias, researchers recently proposed a more complex dual inverse distance weighting technique. In this chapter, we provide a theoretical explanation both for the inverse distance weighting and for the dual inverse distance weighting. Specifically, we show that if we use the general fuzzy ideas to formally describe the desired property of the interpolation procedure, then physically natural scale-invariance requirement select only these two distance weighting techniques.

Comment. Results from this chapter first appeared in [1].

18.1 Formulation of the Problem

Need for interpolation of spatial data. In many practical situations, we are interested in the value of a certain physical quantity at different spatial locations. For example, in geosciences, we may be interested in how elevation and depths of different geological layers depend on the spatial location. In environmental sciences,

L. Bokati and V. Kreinovich, *Decision Making Under Uncertainty, with a Special Emphasis on Geosciences and Education*, Studies in Systems, Decision and Control 218,
https://doi.org/10.1007/978-3-031-26086-5_18

we may be interested in the concentration of different substances in the atmosphere at different locations. etc.

In principle, at each location, we can measure–directly or indirectly–the value of the corresponding quantity. However, we can only perform the measurement at a finite number of locations. Since we are interested in the values of the quantity at all possible locations, we need to estimate these values based on the measurement results–i.e., we need to *interpolate* and *extrapolate* the spatial data.

In precise terms: we know the values $q_i = q(x_i)$ of the quantity of interest q at several locations x_i, $i = 1, 2, \ldots, n$. Based on this information, we would like to estimate the value $q(x)$ of this quantity at a given location x.

Inverse distance weighting. A reasonable estimate q for $q(x)$ is a weighted average of the known values $q(x_i)$: $q = \sum\limits_{i=1}^{n} w_i \cdot q_i$, with $\sum\limits_{i=1}^{n} w_i = 1$. Naturally, the closer is the point x to the point x_i, the larger should be the weight w_i–and if the distance $d(x, x_i)$ is large, then the value $q(x_i)$ should not affect our estimate at all. So, the weight w_i with which we take the value q_i should decrease with the distance.

Empirically, it turns out that the best interpolation is attained when we take the weight proportional to some negative power of the distance: $w_i \sim (d(x, x_i))^{-p}$ for some $p > 0$. Since the weights have to add up to 1, we thus get

$$w_i = \frac{(d(x, x_i))^{-p}}{\sum\limits_{j=1}^{n} (d(x, x_j))^{-p}}.$$

This method–known as *inverse distance weighting*–is one of most widely used spatial interpolation methods; see, e.g., [2–7].

First challenge: why inverse distance weighting? In general, the fact that some algorithm is empirically the best means that we tried many other algorithms, and this particular algorithm worked better than everything else we tried. In practice, we cannot try all possible algorithms, we can only try finitely many different algorithms. So, in principle, there could be an algorithm that we did not try and that will work better than the one which is currently empirically the best.

To be absolutely sure that the empirically found algorithm is the best, it is thus not enough to perform more testing: we need to have some theoretical explanation of this algorithm's superiority. Because of this, every time we have some empirically best alternative, it is desirable to come up with a theoretical explanation of why this alternative is indeed the best–and if such an explanation cannot be found, maybe this alternative is actually not the best.

Thus, the empirical success of inverse distance weighting prompts a natural question: is this indeed the best method? This is the first challenge that we will deal with in this paper.

Limitations of inverse distance weighting. While the inverse distance weighting method is empirically the best among different distance-dependence interpolation techniques, it has limitations; see, e.g., [8].

Specifically, it works well when we have a reasonably uniformly distributed spatial data. The problem is that in many practical cases, we have more measurements in some areas and fewer in others. For example, when we measure meteorological quantities such as temperature, humidity, wind speed, we usually have plenty of sensors (and thus, plenty of measurement results) in cities and other densely populated areas, but much fewer measurements in not so densely populated areas–e.g., in the deserts.

Let us provide a simple example explaining why this may lead to a problem. Suppose that we have two locations A and B at which we perform measurements:

- Location A is densely populated, so we have two measurement results q_A and $q_{A'}$ from this area.
- Location B is a desert, so we have only one measurement result q_B from this location.

Since locations A and A' are very close, the corresponding values are also very close, so we can safely assume that they are equal: $q_A = q_{A'}$. Suppose that we want to use these three measurement results to predict the value of the quantity x at a midpoint C between the locations A and B.

Since C is exactly in the middle between A and B, when estimating q_C, intuitively, we should combine the values q_A and q_B with equal weights, i.e., take $q_C = \frac{q_A + q_B}{2}$. From the commonsense viewpoint, it should not matter whether we made a single measurement at the location A or we made two different measurements.

However, this is *not* what we get if we apply the inverse distance weighting. Indeed, in this case, since all the distance are equal $d(A, C) = d(A', C) = d(B, C)$, the inverse distance weighting leads to

$$q_C = \frac{q_A + q_{A'} + q_B}{3} = \frac{2}{3} \cdot q_A + \frac{1}{3} \cdot q_B.$$

Dual inverse distance weighting: an empirically efficient way to overcome this limitation. To overcome the above limitation, a recent paper [8] proposed a new method called *dual inverse distance weighting*, a method that is empirically better than all previously proposed attempts to overcome this limitation.

In this method, instead of simply using the weight $w_i \sim (d(x, x_i))^{-p}$ depending on the distance, we also give more weight to the points which are more distant from others–and less weight to points which are close to others, by using a formula

$$w_i \sim (d(x, x_i))^{-p} \cdot \left(\sum_{j \neq i} (d(x_i, x_j))^{p_2} \right), \text{ for some } p_2 > 0.$$

Let us show, on an example, that this idea indeed helps overcome the above limitation. Indeed, in the above example of extrapolating from the three points $A \approx A'$

and B to the midpoint C between A and B (for which $d(A, C) = d(B, C)$), we have $d(A, A') \approx 0$ and $d(A, B) \approx d(A', B)$. Thus, we get the following expressions for the additional factors $f_i = \sum_{j \neq i} (d(x_i, x_j))^{p_2}$:

$$f_A = (d(A, A'))^{p_2} + (d(A, B))^{p_2} \approx (d(A, B))^{p_2},$$

$$f_{A'} = (d(A', A))^{p_2} + (d(A', B))^{p_2} \approx (d(A, B))^{p_2},$$

and

$$f_B = (d(B, A))^{p_2} + (d(B, A'))^{p_2} \approx 2(d(A, B))^{p_2}.$$

So, the weights w_A and $w_{A'}$ with which we take the values q_A and $q_{A'}$ are proportional to

$$w_A \approx w_{A'} \sim (d(A, C))^{-p} \cdot f_A \approx (d(A, C))^{-p} \cdot (d(A, B))^{p_2},$$

while

$$w_B \approx w_B \sim (d(B, C))^{-p} \cdot f_2 \approx (d(A, C))^{-p} \cdot 2(d(A, B))^{p_2}.$$

The weight w_B is thus twice larger than the weights w_A and $w_{A'}$: $w_B = 2w_A = 2w_{A'}$. So the interpolated value of q_C is equal to

$$q_C = \frac{w_A \cdot q_A + w_{A'} \cdot q_{A'} + w_B \cdot q_B}{w_A + w_{A'} + w_B} = \frac{w_A \cdot q_A + w_A \cdot q_{A'} + 2w_A \cdot q_A}{w_A + w_{A'} + 2w_A}.$$

Dividing both numerator and denominator by $2w_A$ and taking into account that $q_{A'} = q_A$, we conclude that $q_C = \dfrac{q_A + q_B}{2}$, i.e., exactly the value that we wanted.

Second challenge: why dual inverse distance weighting? In view of the above, it is also desirable to come up with a theoretical explanation for the dual inverse weighting method as well. This is the second challenge that we take on in this paper.

18.2 What Is Scale Invariance and How It Explains the Empirical Success of Inverse Distance Weighting

What is scale invariance. When we process the values of physical quantities, we process real numbers. It is important to take into account, however, that the numerical value of each quantity depends on the measuring unit. For example, suppose that we measure the distance in kilometers and get a numerical value d such as 2 km. Alternatively, we could use meters instead of kilometers. In this case, the exact same distance will be described by a different number: 2000 m.

In general, if we replace the original measuring unit with a new one which is λ times smaller, all numerical values will be multiplied by λ, i.e., instead of the original numerical value x, we will get a new numerical value $\lambda \cdot x$.

Scale-invariance means, in our case, that the result of interpolation should not change if we simply change the measuring unit. Let us analyze how this natural requirement affects interpolation.

General case of distance-dependent interpolation. Let us consider the general case, when the further the point, the smaller the weight, i.e., in precise terms, when the weight w_i is proportional to $f(d(x, x_i))$ for some decreasing function $f(z)$: $w_i \sim f(d(x, x_i))$. Since the weights should add up to 1, we conclude that

$$w_i = \frac{f(d(x, x_i))}{\sum\limits_j f(d(x, x_j))}, \tag{18.1}$$

and thus, our estimate q for $q(x)$ should take the form

$$q = \sum_{i=1}^{n} \frac{f(d(x, x_i))}{\sum\limits_j f(d(x, x_j))} \cdot q_i. \tag{18.2}$$

In this case, scale-invariance means that for each $\lambda > 0$, if we replace all the numerical distance values $d(x, x_i)$ with "re-scaled" values $\lambda \cdot d(x, x_i)$, then we should get the exact same interpolation result, i.e., that for all possible values of q_i and $d(x, x_i)$, we should have

$$\sum_{i=1}^{n} \frac{f(\lambda \cdot d(x, x_i))}{\sum\limits_j f(\lambda \cdot d(x, x_j))} \cdot q_i = \sum_{i=1}^{n} \frac{f(d(x, x_i))}{\sum\limits_j f(d(x, x_j))} \cdot q_i. \tag{18.3}$$

Scale-invariance leads to inverse distance weighting. Let us show that the requirement (18.3) indeed leads to inverse distance weighting.

Indeed, let us consider the case when we have only two measurement results:

- at the point x_1, we got the value $q_1 = 1$, and
- at point x_2, we got the value $q_2 = 0$.

Then, for any point x, if we use the original distance values $d_1 \stackrel{\text{def}}{=} d(x, x_1)$ and $d_2 \stackrel{\text{def}}{=} d(x, x_2)$, the interpolated value q at this point will have the form

$$q = \frac{f(d_1)}{f(d_1) + f(d_2)}.$$

On the other hand, if we use a λ times smaller measuring unit, then the extrapolation formula leads to the values

$$\frac{f(\lambda \cdot d_1)}{f(\lambda \cdot d_1) + f(\lambda \cdot d_2)}.$$

The requirement that the interpolation value does not change if we simply change the measuring unit implies that these two expression must coincide, i.e., that we must have:

$$\frac{f(\lambda \cdot d_1)}{f(\lambda \cdot d_1) + f(\lambda \cdot d_2)} = \frac{f(d_1)}{f(d_1) + f(d_2)}. \tag{18.4}$$

If we take the inverse of both sides of this formula, i.e., flip the numerator and denominator in both sides, we get

$$\frac{f(\lambda \cdot d_1) + f(\lambda \cdot d_2)}{f(\lambda \cdot d_1)} = \frac{f(d_1) + f(d_2)}{f(d_1)}. \tag{18.5}$$

Subtracting number 1 from both sides, we get a simplified expression

$$\frac{f(\lambda \cdot d_2)}{f(\lambda \cdot d_1)} = \frac{f(d_2)}{f(d_1)}. \tag{18.6}$$

If we divide both sides by $f(d_2)$ and multiply by $f(\lambda \cdot d_1)$, we get the equivalent equality in which variables d_1 and d_2 are separated:

$$\frac{f(\lambda \cdot d_2)}{f(d_2)} = \frac{f(\lambda \cdot d_1)}{f(d_1)}. \tag{18.7}$$

The left-hand side of this formula does not depend on d_1; thus, the right-hand side does not depend on d_1 either, it must thus depend only on λ. Let us denote this right-hand side by $c(\lambda)$. Then, from $\dfrac{f(\lambda \cdot d_1)}{f(d_1)} = c(\lambda)$, we conclude that

$$f(\lambda \cdot d_1) = c(\lambda) \cdot f(d_1) \tag{18.8}$$

for all possible values of $\lambda > 0$ and d_1.

It is known that for decreasing functions $f(z)$, the only solutions to the functional equation (18.8) are functions $f(z) = c \cdot z^{-p}$ for some $p > 0$; see, e.g., [9]. For this function $f(z)$, the extrapolated value has the form $\sum f_i \cdot q_i$, with

$$f_i = \frac{c \cdot (d(x, x_i))^{-p}}{\sum\limits_{j=1}^{n} c \cdot (d(x, x_j))^{-p}}.$$

If we divide both numerator and denominator by c, we get exactly the inverse distance weighting formula.

Thus, scale-invariance indeed leads to inverse distance weighting.

Comment. For smooth function $f(x)$, the above result about solutions of the functional equation can be easily derived. Indeed, differentiating both sides of the equality (8) by λ and taking $\lambda = 1$, we get

$$f'(d_1) \cdot d_1 = \alpha \cdot f(d_1),$$

where we denoted $\alpha \stackrel{\text{def}}{=} c'(1)$, i.e., we have

$$\frac{df}{dd_1} = \alpha \cdot f.$$

If we divide both sides by f and multiply by dd_1, we separate d_1 and f: $\dfrac{df}{f} = \alpha \cdot \dfrac{dd_1}{d_1}$. Integrating both sides, we get $\ln(f) = \alpha \cdot \ln(d_1) + C$, where C is the integration constant. Applying $\exp(z)$ to both sides and taking into account that $\exp(\ln(f)) = f$ and

$$\exp(\alpha \cdot \ln(d_1) + C) = \exp(\alpha \cdot \ln(d_1)) \cdot \exp(C) = \exp(C) \cdot (\exp(\ln(d_1))^{\alpha} = \exp(C) \cdot d_1^{\alpha},$$

we get $f(d_1) = c \cdot d_1^{\alpha}$, where we denoted $c \stackrel{\text{def}}{=} \exp(C)$. Since the function $f(z)$ is decreasing, we should have $\alpha < 0$, i.e., $\alpha = -p$ for some $p > 0$. The statement is proven.

18.3 Scale Invariance and Fuzzy Techniques Explain Dual Inverse Distance Weighting

What we want: informal description. In the previous section, when computing the estimate q for the value $q(x)$ of the desired quantity at a location x, we used, in effect, the weighted average of the measurements results q_i, with the weights decreasing as the distance $d(x, x_i)$ increases–i.e., in more precise terms, with weights proportional to $f(d(x, x_i))$ for some decreasing function $f(z)$. In this case, scale-invariance implies that $f(z) = z^{-p}$ for some $p > 0$.

As we have mentioned in Sect. 18.1, we need to also give more weight to measurements at locations x_i which are far away from other location–and, correspondingly, less weight to measurements at locations which are close to other locations. In terms of weights, we would like to multiply the previous weights $f(d(x, x_i)) = (d(x, x_i))^{-p}$ by an additional factor f_i depending on how far away is location x_i from other locations. The further away the location x_i from other locations, the higher the factor f_i shall be. In other words, the factor f_i should be larger or smaller depending on our degree of confidence in the following statement:

$d(x_i, x_1)$ is large and $d(x_i, x_2)$ is large and ... $d(x_i, x_n)$ is large.

Let us use fuzzy techniques to translate this informal statements into precise terms. To translate the above informal statement into precise terms, a reasonable idea is to use fuzzy techniques–techniques specifically designed for such a translation; see, e.g., [10–15]. In this technique, to each basic statement–like "d is large"–we assign a degree to which, according to the expert, this statement is true. This degree is usually denoted by $\mu(d)$. In terms of these notations:

- the degree to which $d(x_i, x_1)$ is large is equal to $\mu(d(x_i, x_1))$;
- the degree to which $d(x_i, x_2)$ is large is equal to $\mu(d(x_i, x_2))$; etc.

To estimate the degree to which the above "and"-statement is satisfied, fuzzy techniques suggest that we combine the above degrees by using an appropriate "and"-operation (= t-norm) $f_\&(a, b)$. Thus, we get the following degree:

$$f_\&(\mu(d(x_i, x_1)), \mu(d(x_i, x_2)), \ldots, \mu(d(x_i, x_{i-1})), \mu(d(x_i, x_{i+1})), \ldots, \mu(d(x_i, x_n))).$$

It is known–see, e.g., [16]–that for any "and"-operation and for any $\varepsilon > 0$, there exists an ε-close "and"-operation of the type $f_\&(a, b) = g^{-1}(g(a) + g(b))$ for some monotonic function $g(a)$, where $g^{-1}(a)$ denotes the inverse function (i.e., the function for which $g^{-1}(a) = b$ if and only if $g(b) = a$). Since the approximation error ε can be arbitrarily small, for all practical purposes, we can safely assume that the actual "and"-operation has this g-based form. Substituting this expression for the "and"-operation into the above formula, we conclude that f_i should monotonically depend on the expression

$$g^{-1}(g(\mu(d(x_i, x_1))) + \cdots + g(\mu(d(x_i, x_n)))).$$

Since the function g^{-1} is monotonic, this means that f_i is a monotonic function of the expression

$$G(d(x_i, x_1)) + \cdots + G(d(x_i, x_n))),$$

where we denoted $G(d) \stackrel{\text{def}}{=} g(\mu(d))$. In other words, we conclude that

$$f_i = F(G(d(x_i, x_1)) + \cdots + G(d(x_i, x_n))) \tag{18.9}$$

for some monotonic function $F(z)$.

So, we get an estimate

$$q = \sum_{i=1}^{n} \frac{f_i \cdot (d(x, x_i))^{-p} \cdot q_i}{\sum_{j=1}^{n} f_j \cdot (d(x, x_j))^{-p}}, \tag{18.10}$$

where the factors f_i are described by the formula (18.9).

Let us recall the motivation for the factors f_i. As we have mentioned earlier, the main motivation for introducing the factors f_i is to make sure that for the midpoint C between A and B, we will have the estimate $\frac{q_A + q_B}{2}$, even if we perform two (or more) measurements at the point A. Let us analyze for which functions $F(z)$ and $G(z)$ this requirement is satisfied.

For the purpose of this analysis, let us consider the case when we have m measurement locations $A_1, \ldots, A_m$ in the close vicinity of the location A and one measurement result at location B. Let d denote the distance $d(A, B)$ between the locations A and B. For all the measurement locations $A_1, \ldots, A_m$, and B, the distance to the point C is the same–equal to $d/2$. Thus, in this case, the factors $(d(x, x_i))^{-p}$ in the formula (18.10) are all equal to each other. So, we can divide both the numerator and the denominator by the formula (18.10) by this common factor, and get a simplified expression

$$q = \sum_{i=1}^{n} \frac{f_i \cdot q_i}{\sum_{j=1}^{n} f_j}.$$

Since for the points $A_1, \ldots, A_m$ we have the same measurement results q_i (we will denote them by q_A), and the same factors f_i (we will denote them by f_A), we get

$$q = \frac{m \cdot f_A \cdot q_A + f_B \cdot q_B}{m \cdot f_A + f_B}. \tag{18.11}$$

We want to make sure that this value is equal to the arithmetic average $\frac{q_A + q_B}{2}$. Thus, the coefficient at q_A in the formula (18.11) should be equal to $1/2$:

$$\frac{m \cdot f_A}{m \cdot f_A + f_B} = \frac{1}{2}.$$

If we multiply both side by their denominators and subtract $m \cdot f_A$ from both sides, we get $m \cdot f_A = f_B$. Due to the formula (18.9), this means

$$m \cdot F(G(d) + (m - 1) \cdot G(0)) = F(m \cdot G(d)). \tag{18.12}$$

In the limit $d = 0$, this formula becomes $m \cdot F(m \cdot G(0)) = F(m \cdot G(0))$, thus $F(m \cdot G(0)) = 0$. Since the function $F(z)$ is monotonic, we cannot have $G(0) \neq 0$, since then we would have $F(z) = 0$ for all z. Thus, $G(0) = 0$, $F(G(0)) = F(0) = 0$, and the formula (18.12) takes the form $F(m \cdot G(d)) = m \cdot F(G(d))$. This is true for any value $z = G(d)$, so we have $F(m \cdot z) = m \cdot F(z)$ for all m and z.

- In particular, for $z = 1$, we get $F(m) = c \cdot m$, where $c \stackrel{\text{def}}{=} F(1)$.
- For $z = 1/m$, we then have $F(1) = c = m \cdot F(1/m)$, hence

$$F(1/m) = c \cdot (1/m).$$

- Similarly, we get $F(p/q) = F(p \cdot (1/q)) = p \cdot F(1/q) = p \cdot (c \cdot (1/q)) = c \cdot (p/q)$. So, for all rational values $z = p/q$, we get $F(z) = c \cdot z$.

Since the function $F(z)$ is monotonic, the formula $F(z) = c \cdot z$ is true for all values z.

Dividing both the numerator and the denominator by the coefficient c, we conclude that

$$q = \sum_{i=1}^{n} \frac{F_i \cdot (d(x, x_i))^{-p} \cdot q_i}{\sum_{j=1}^{n} F_j \cdot (d(x, x_j))^{-p}}, \tag{18.13}$$

where we denoted

$$F_i \stackrel{\text{def}}{=} G(d(x_i, x_1)) + \cdots + G(d(x_i, x_n)). \tag{18.14}$$

Let us now use scale-invariance. We want to make sure that the estimate (18.13) does not change after re-scaling $d(x, y) \to d'(x, y) = \lambda \cdot d(x, y)$, i.e., that the same value q should be also equal to

$$q = \sum_{i=1}^{n} \frac{F'_i \cdot (d'(x, x_i))^{-p} \cdot q_i}{\sum_{j=1}^{n} F'_j \cdot (d'(x, x_j))^{-p}}, \tag{18.15}$$

where

$$F'_i = G(d'(x_i, x_1)) + \cdots + G(d'(x_i, x_n)). \tag{18.16}$$

Here, $(d'(x, x_i))^{-p} = \lambda^{-p} \cdot (d(x, x_i))^{-p}$. Dividing both the numerator and the denominator of the right-hand side of the formula (18.15) by λ^{-p}, we get a simplified expression

$$q = \sum_{i=1}^{n} \frac{F'_i \cdot (d(x, x_i))^{-p} \cdot q_i}{\sum_{j=1}^{n} F'_j \cdot (d(x, x_j))^{-p}}. \tag{18.17}$$

The two expressions (18.13) and (18.17) are linear in q_i. Thus, their equality implies that coefficients at each q_i must be the same. In particular, this means that the ratios of the coefficients at q_1 and q_2 must be equal, i.e., we must have

$$\frac{F_1 \cdot (d(x, x_1))^{-p}}{F_2 \cdot (d(x, x_2))^{-p}} = \frac{F_1' \cdot (d(x, x_1))^{-p}}{F_2' \cdot (d(x, x_2))^{-p}},$$

i.e.,

$$\frac{F_1}{F_2} = \frac{F_1'}{F_2'}.$$

For the case when we have three points with $d(x_1, x_2) = d(x_1, x_3) = d$ and $d(x_2, x_3) = D$, due to the formula (18.14), this means that

$$\frac{2G(d)}{G(d) + G(D)} = \frac{2G(\lambda \cdot d)}{G(\lambda \cdot d) + G(\lambda \cdot D)}.$$

Inverting both sides, multiplying both sides by 2 and subtracting 1 from both sides, we conclude that

$$\frac{G(D)}{G(d)} = \frac{G(\lambda \cdot D)}{G(\lambda \cdot d)}$$

for all λ, d, and D. We already know–from the first proof–that this implies that $G(d) = c \cdot d^{p_2}$ for some c and p_2, and that, by dividing both numerator and denominator by c, we can get $c = 1$.

Thus, we indeed get a justification for the dual inverse distance weighting.

References

1. L. Bokati, A. Velasco, V. Kreinovich, Scale-invariance and fuzzy techniques explain the empirical success of inverse distance weighting and of dual inverse distance weighting in geosciences, in *Proceedings of the Annual Conference of the North American Fuzzy Information Processing Society NAFIPS'2020*, Redmond, Washington, August 20–22, (2020), pp. 379–390
2. Q. Chen, G. Liu, X. Ma, G. Marietoz, Z. He, Y. Tian, Z. Weng, Local curvature entropy-based 3D terrain representation using a comprehensive quadtree. ISPRS J. Photogramm. Remote. Sens. **139**, 130–145 (2018)
3. K.C. Clarke, *Analytical and Computer Cartography* (Pnetice Hall, Englewood Cliffs, New Jersey, 1990)
4. N. Henderson, L. Pena, The inverse distance weighting interpolation applied to a particular form of the path rubes method: theory and computation for advection in uncompressible flow. Appl. Math. Comput. **304**, 114–135 (2017)
5. Q. Liang, S. Nittel, J.C. Whittier, S. Bruin, Real-time inverse distance weighting interpolation for streaming sensor data. Trans. GIS. **22**(5), 1179–1204 (2018)
6. I. Loghmari, Y. Timoumi, A. Messadi, Performance comparison of two global solar radiation models for spatial interpolation purposes. Renew. Sustain. Energy Rev. **82**, 837–844 (2018)
7. D. Shepard, A two-dimensional interpolation function for irregularly-spaced data, in *Proceedings of the 1968 23rd ACM National Conference*, (1968), pp. 517–524

8. Z. Li, X. Zhang, R. Zhu, Z. Zhiang, Z. Weng, Integrating data-to-data correlation into inverse distance weighting. Comput. Geosci. (2019) https://doi.org/10.1007/s10596-019-09913-9
9. J. Aczel, J. Dhombres, *Functional Equations in Several Variables* (Cambridge University Press, Cambridge, UK, 1989)
10. R. Belohlavek, J.W. Dauben, G.J. Klir, *Fuzzy Logic and Mathematics: A Historical Perspective* (Oxford University Press, New York, 2017)
11. G. Klir, B. Yuan, *Fuzzy Sets and Fuzzy Logic* (Prentice Hall, Upper Saddle River, New Jersey, 1995)
12. J.M. Mendel, *Uncertain Rule-Based Fuzzy Systems: Introduction and New Directions* (Springer, Cham, Switzerland, 2017)
13. H.T. Nguyen, C.L. Walker, E.A. Walker, *A First Course in Fuzzy Logic* (Chapman and Hall/CRC, Boca Raton, Florida, 2019)
14. V. Novák, I. Perfilieva, J. Močkoř, *Mathematical Principles of Fuzzy Logic* (Kluwer, Boston, Dordrecht, 1999)
15. L.A. Zadeh, Fuzzy sets. Inf. Control. **8**, 338–353 (1965)
16. H.T. Nguyen, V. Kreinovich, P. Wojciechowski, Strict archimedean t-norms and t-conorms as universal approximators. Int. J. Approx. Reason. **18**(3–4), 239–249 (1998)

Chapter 19
Dynamic Triggering of Earthquakes

In the previous chapter, we analyzed the best ways to extend our knowledge to other spatial locations. In this chapter, we analyze how to extend it to the future moments of time. As an example of extending knowledge to future moments of time–i.e., prediction–we deal with one of the least studied seismic phenomena: earthquakes triggering other earthquakes.

There are two empirical phenomena associated with this triggering, these phenomena are related to the orientation of the triggering wave and to the distance from the original earthquake.

First, it is known that seismic waves from a remote earthquake can trigger a small local earthquake. Recent analysis has shown that this triggering occurs mostly when the direction of the incoming wave is orthogonal to the direction of the local fault, some triggerings occur when these directions are parallel, and very few triggerings occur when the angle between these two directions is different from 0 and 90°. In the first part of this chapter, we propose a symmetry-based geometric explanation for this unexpected observation.

The second phenomenon is that while some of the triggered earthquakes are strong themselves, strong triggered earthquakes only happen within a reasonably small distance (less than 1000 km) from the original earthquake. Even catastrophic earthquakes do not trigger any strong earthquakes beyond this distance. In the second part of this chapter, we provide a possible geometric explanation for this phenomenon.

Comment. Results from this chapter first appeared in [1, 2].

L. Bokati and V. Kreinovich, *Decision Making Under Uncertainty, with a Special Emphasis on Geosciences and Education*, Studies in Systems, Decision and Control 218,
https://doi.org/10.1007/978-3-031-26086-5_19

19.1 Formulation of the First Problem

Dynamic triggering of earthquakes: a phenomenon. When a seismic wave from an earthquake hits a distant fault–completely unrelated to the fault involved in the original earthquake–this sometimes triggers a minor earthquake and/or other seismic activity at the distant fault's location–either almost immediately, or after some delay; see, e.g., [3–13] and references therein.

Somewhat unexpected feature of dynamic triggering. Interestingly, it turned out that triggering strongly depends on the angle between the direction of the incoming wave and the direction of the fault; see, e.g., [5, 14]. The vast majority of triggerings occur when these direction are (almost) orthogonal to each other. There is another spike of triggerings–mush smaller one–when the direction of the incoming wave is practically parallel to the direction of the fault. Very few triggerings occur when the angle is different from 0 and from 90°.

Why this happens is not clear.

What we do in this chapter. In this chapter, we provide a symmetry-based geometric explanation for the above feature of dynamic triggering.

19.2 Symmetry-Based Geometric Explanation

Let us use symmetries. In physics, symmetries–in particular, geometric symmetries–are an important tool that helps analyze and explain many physical phenomena; see, e.g., [15, 16]. In view of this, let us consider geometric symmetries related to dynamic triggering of earthquakes.

Symmetries: no-faults case. In an area without faults, all physical properties are the same at different locations and at different directions. So, if we shift in any direction, or rotate, the situation remains the same. In addition to these continuous symmetries, we can also consider discrete symmetries:

- reflections over any line and
- reflections in any point.

Also, many physical properties do not change if we re-scale the area, i.e., change the original coordinates (x, y) by re-scaled values $(\lambda \cdot x, \lambda \cdot y)$, for some $\lambda > 0$.

Which symmetries remain in the presence of the fault? Locally, most faults are straight lines.

Thus, when there is a fault, the resulting configuration is no longer invariant with respect to all the above geometric transformations. For example, there is no longer invariance with respect to rotations, since rotating the configuration will also rotate the direction of the fault–and thus, make the configuration different.

The only remaining symmetries are:

- shifts in the direction of the fault $x \to x + a$;
- reflection over the fault line;
- reflections over any line orthogonal to the fault; and
- scalings $(x, y) \to (\lambda \cdot x, \lambda \cdot y)$ and reflections in any point on the fault.

This configuration and its symmetries are described in the picture below:

- shifts in the direction of the fault are marked by a horizontal one-directional arrow;
- reflection over the fault line is marked by a vertical bi-directions arrow;
- a line orthogonal to the fault is marked as a dashed line, and the reflection over this line is marked by a horizontal bi-directions arrow; and
- scalings and the reflection in a point on the fault are marked by a slanted bi-directional arrow.

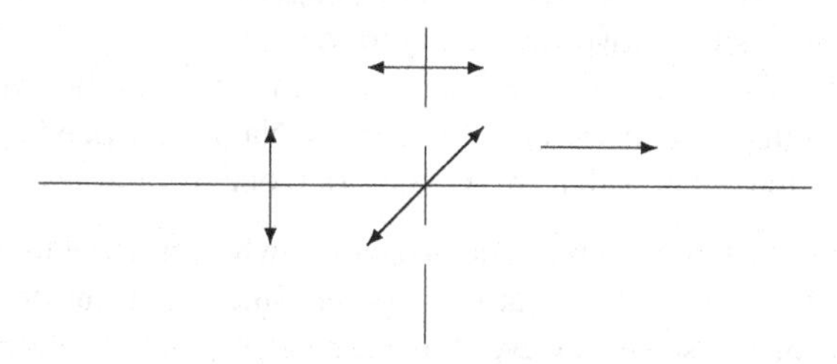

Symmetry matching and resonance: general reminder. To describe how symmetries influence the physical effect, let us consider a simple example: a pendulum with period T. The periodicity of a pendulum means that if we shift the time by T, i.e., consider a moment $t + T$ instead of the original moment t, the state of the pendulum will remain the same. Similarly, the pendulum remains invariant if we shift all the moments by time $k \cdot T$, for some integer k.

If we perturb the pendulum at random moments of time, it will be affected, but overall, not much: randomly applied pushes will cancel each other, and the overall effect will be small. The largest possible effect can be obtained if we apply perturbations that have the exact same symmetry–i.e., if at the same moment of time during each period, we apply the exact same small push. This is how the kids play on the swings.

For the pendulum and other time-shifting situations, this phenomenon is called a *resonance*, but the same phenomenon occurs in other situations as well, when symmetries are not necessarily related to time: e.g., the largest effect of a wave on a crystal is when the wave's spatial symmetry is correlated with the symmetry of the crystal.

If some symmetries are preserved, we still get some effect: e.g., we can still affect the swings if we push at every other cycle (thus keeping it only invariant with respect to shifts by $2T$), or at every third cycle, etc.

In all these cases, the more symmetries are preserved, the larger the effect; see, e.g., [15, 16].

Symmetries of a configuration involving a seismic wave. From this viewpoint, let us consider the dynamic triggering of earthquakes. The seismic wave generated by a remote earthquake usually lasts for a short time. So, at any given moment of time, what we see is the front line corresponding to the current location of the seismic wave. Locally this line is also a straight line.

The triggering happens when the seismic wave affects the fault, i.e., when the two lines intersect. Thus, from the geometric viewpoint, we need to consider a configuration in which we have two intersecting lines:

- the line corresponding to the fault, and
- the line corresponding to the current position of the seismic wave.

As we have mentioned, the more symmetries are preserved in comparison with the original configuration in which there is only the fault (and no seismic waves), the stronger will be the effect (in this case, the triggering effect). So, to find out which configurations of the two lines lead to a larger effect, we need to describe the symmetries of the resulting two-line configuration.

A simple geometric analysis shows that the symmetries of the two-line configuration depend on the angle between the two lines. Namely, depending on the angle, we have three different cases that we will describe one by one.

Case when the two lines coincide. The first case is when the front line of the seismic wave coincides with the fault line at some point. Since the front line is orthogonal to the direction of the seismic wave, this case corresponds to the case when the direction of the seismic wave is orthogonal to the fault. In this case, the two-line configuration simply coincides with the fault-only configuration. Thus, in this case, all the symmetries of the fault-only configuration are preserved.

This is thus the case when the largest number of symmetries are preserved–thus, the case when we expect the strongest triggering effect. This is indeed what we observe.

Case when the two lines are orthogonal. The front line of a seismic wave is orthogonal when the direction of the seismic wave is parallel to the direction of the fault.

In this case, we no longer have invariance with respect to shifts in the direction of the fault. However, we still have invariance with respect to:

- scaling $(x, y) \to (\lambda \cdot x, \lambda \cdot y)$,
- a reflection in an intersection point $(x, y) \to (-x, -y)$, and
- reflections over each of the two lines.

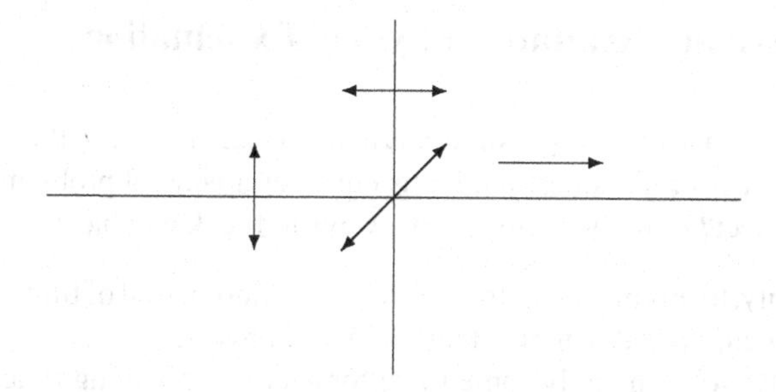

Remaining case when the two lines are neither parallel not orthogonal. In this case, the only remaining symmetries are:

- scalings and
- reflection in an intersection point $(x, y) \to (-x, -y)$.

We no longer have invariance with respect to reflection over any line.

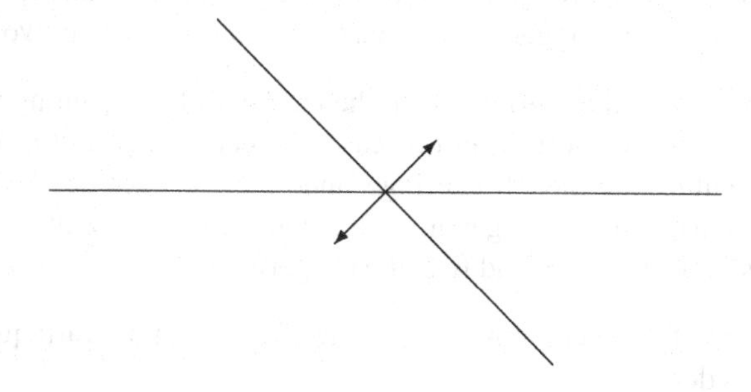

Conclusion. Our analysis shows that:

- the largest number of symmetries are preserved when the direction of the seismic wave is orthogonal to the fault;
- somewhat fewer symmetries are preserved when the direction of the seismic wave is parallel to the fault;
- and in other configurations, we have the smallest number of preserved symmetries.

In line with the above general physical analysis, we expect that:

- the most triggering effects will happen when the seismic wave is orthogonal to the fault;
- somewhat fewer triggering effects will occur when the seismic wave is parallel to the fault; and
- the smallest number of triggering effects will occur when the seismic wave is neither parallel not orthogonal to the fault.

This is exactly what we observe. Thus, the symmetry-based geometric analysis indeed explains the observed relative frequency of dynamic triggering of earthquakes at different angles.

19.3 A Possible Qualitative Physical Explanation

General idea. What are the possible *physical* explanations for the observed phenomenon? To answer this question, let us consider a general problem: how do you break some object? Usually, there are two ways to break an object:

- You can apply, to this object, a strong force for a short period of time. This happens, e.g., when a cup falls down on a hard floor and breaks.
- Alternatively, you can apply some force for a sufficiently long time. This is what happens to structures under stress: they eventually start to crumble.

The first type of breaking is most assured: a rare china cup can withstand a fall. The second type of breaking is not guaranteed: many old building stand for hundreds of years without breaking down, but it still occurs–some buildings do eventually collapse if they are not well maintained.

Analysis of the problem. From this viewpoint, to describe when an incoming seismic wave is most probable to trigger an earthquake, we should look at two situations:

- a situation when the time during which the energy of the incoming seismic wave affects the fault is the shortest; in this case, the energy per unit time will be the largest–this will lead to most triggerings, and
- a situation when the time during which the incoming seismic wave affects the fault is the longest–this will also lead to some triggerings.

The wavefront of the incoming seismic wave from a remote earthquake is practically flat. Let us denote:

- the fault length by L,
- the speed of the incoming seismic wave by v,
- the angle between the direction of the wave and the fault by α, and
- the time during which this wave affects the fault by t.

The wavefront is orthogonal to the direction in which the wave comes, so the angle between the wavefront and the fault is $90 - \alpha$. Let us consider the moment when the seismic wave first hits one of the sides of the fault. Let us denote this point by A, and the other side of the fault by B. Eventually, the wavefront will hit the point B as well. We can trace this hit by placing a line from B to the current position of the wavefront along the direction of the incoming wave. Let us denote the point where this line intersects with the current wavefront by C; this point marks the part of the seismic wave that, in time t, will hit the point B on the fault. The wave travels with speed v, so the distance BC is equal to $v \cdot t$. In the right triangle ACB, the angle $\angle BAC$ is equal to $90 - \alpha$, thus $\angle ABC = \alpha$, and thus, by definition of cosine, $v \cdot t = L \cdot \cos(\alpha)$. Hence, the interaction time is equal to

$$t = \frac{L \cdot \cos(\alpha)}{v}.$$

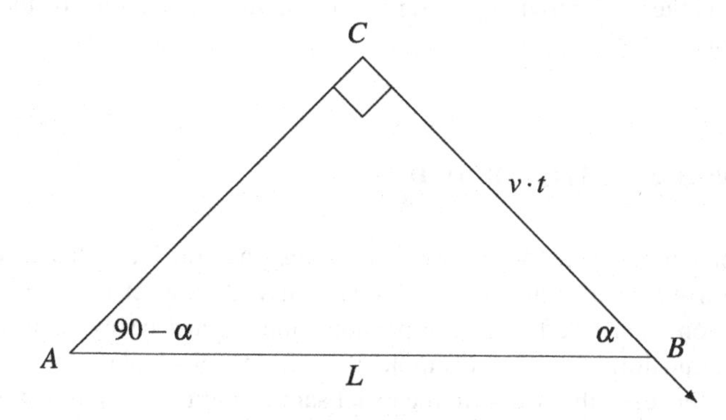

Resulting explanation. This time is the smallest when $\cos(\alpha)$ is the smallest, i.e., when α is close to 90° (when the cosine is 0). This is indeed when we observe most triggerings.

The interaction time t is the largest when the cosine takes the largest value, i.e., when the angle α is close to 0 (when the cosine is 1). This is indeed when we observe the second (smaller) peak of triggerings.

So, indeed, we seem to have–at least on a qualitative level–a possible physical explanation for the observed phenomenon.

Remaining problem. So far, both the symmetry-based geometric analysis and the physical analysis provide us only with a qualitative explanation of the observed phenomenon. It is desirable to transform this qualitative explanation into a quantitative one–e.g., to be able to predict which proportion of triggerings occurs at different angles.

19.4 Formulation of the Second Problem

Triggered earthquakes: original expectations. It is known that seismic waves from a large earthquake can trigger earthquakes at some distance from the original quake; see, e.g., [17–24]. At first glance, it seems reasonable to conclude that the stronger the original earthquake, the stronger will be the triggered earthquakes, so that catastrophic earthquakes will trigger strong earthquakes even far away from the original location.

Unexpected empirical fact. Somewhat surprisingly, it turned out that no matter how strong the original earthquake, strong triggered earthquakes are limited to an about 1000 km distance from the original event. At larger distances, the triggered (secondary) earthquakes are all low-magnitude, with magnitude $M < 5$ on Richter scale; see, e.g., [25].

Why? At present, there is no convincing explanation for this empirical fact.

In this chapter, we provide a possible geometric explanation for the observed phenomenon.

19.5 Geometric Explanation

Main idea. Our explanation is based on a very natural idea: that if we have a phenomenon which is symmetric–i.e., invariant with respect to some reasonable transformation–then the effects of this phenomenon will also be invariant with respect to the same transformation. For example, if we have a plank placed, in a symmetric way, over a fence–so that we have the exact same length to the left and to the right of the fence, and we apply similar forces to the left and right ends of this plank, we expect it to curve the same way to the left and to the right of the fence.

What are reasonable transformations here? All related physical processes do not change if we simply shift from one place to another and/or rotate the corresponding configuration by some angle. If we describe each point x by its coordinates x_i, then a shift means that each coordinate x_i is replaced by a shifted value $x_i' = x_i + a_i$, and rotation means that we replace the original coordinates x_i with rotated ones $x_i' = \sum\limits_{j=1}^{n} r_{ij} \cdot x_j$ for an appropriate rotation matrix r_{ij}.

In addition, many physical processes–like electromagnetic or gravitational forces–do not have a fixed spatial scale. If we scale down or scale up, we get the same physical phenomenon (of course, we need to be careful when scaling down or scaling up). This is how, e.g., airplanes were tested before computer simulations were possible: you test a scaled-down model of a plane in a wind tunnel, and it provides a very accurate description of what will happen to the actual airplane. So, to shift and rotation, it is reasonable to add scaling $x_i \to \lambda \cdot x_i$, for an appropriate value λ.

What is the symmetry of the propagating seismic wave? In a reasonable first approximation, the seismic waves propagates equally in all directions with approximately the same speed. So, in this approximation, at any given moment of time, the locations reached by a wave form a circle with radius r equal to the propagation speed times the time from the original earthquake.

When we are close to the earthquake location, we can easily see that the set of all these locations is not a straight line segment, it is a curved part of a circle. However, as we get further and further away from the original earthquake location, this curving becomes less and less visible–just like we easily notice the curvature of a ball, but it is difficult to notice the curvature of an Earth surface; for most experiments, it is safe to assume that locally, the Earth is flat (and this is what people believed for a long time, until more sophisticated measurements showed that it is not flat). So:

- in places close to the original earthquake, the set of locations affected by the incoming seismic wave can be approximated as a circle's arc–a local part of a circle, while
- in places far away from the original earthquake, the set of locations affected by the incoming seismic wave can be well approximated by a straight line segment.

It is important to emphasize that the difference between these two situations depends only on the distance to the original earthquake location, it does not depend on the strength of the earthquake–it is the same for very weak and for very strong earthquakes.

What is the effect of these two different symmetries? Out of all possible symmetries–shifts, rotations, and scalings–a circle is only invariant with respect to all possible rotations around its center. Thus, we expect the effect of the resulting seismic wave to be also invariant with respect to such rotations. Thus, the area A affected by the incoming wave should also be similarly invariant. This means that with each point a, this area must contain the whole circle. As a result, this area consists of one or several such circles. From the viewpoint of this invariance, it could be that the affected area is limited to the circle itself–in which case the area is small, and its effect is small. It can also be that the area includes many concentric circles–in which case the affected area may be significant, and its effect may be significant.

On the other hand, a straight line has different symmetries: it is invariant with respect to shifts along this line and arbitrary scalings. Thus, it is reasonable to conclude that the area effected by such almost-straight-line seismic wave is also invariant with respect to the same symmetries. This implies that this area is limited to the line itself: otherwise, if the area A had at least one point outside the line, then:

- by shifting along the original line, we can form a whole line parallel to the original line, and then
- by applying different scalings, we would get all the lines parallel to the original line–no matter what distance, and thus, we will get the whole plane, while the affected area has to be bounded.

Thus, in such situations, the effect of the seismic wave is limited to the line itself–i.e., in effect, to a narrow area around this line–and will, thus, be reasonably weak.

This indeed explains the absence of remotely triggered large earthquakes. Indeed, for locations close to the earthquake, the resulting phenomenon is (approximately) invariant with respect to rotations–and thus, its effect should be similarly invariant. This leaves open the possibility that a large area will be affected and thus, that the resulting effect will be strong–which explains why in a small vicinity, it is possible to have a triggered large earthquake.

On the other hand, in remote locations, location far away from the original earthquake, the resulting phenomenon is invariant with respect to shifts and scalings–and thus, its effect should be similarly invariant. As a result, only a very small area is affected–which explains why, no matter how strong the original earthquake, it never triggers a large earthquake in such remote locations.

Comments.

- It should be mentioned that our analysis is about the geometric shape of the *area* affected by the seismic wave, not about the physical properties of the seismic wave itself. From the physical viewpoint, at each sensor location, the seismic wave can definitely be treated as a planar wave already at much shorter distances from the original earthquake than 1000 km. However, if instead of limiting ourselves to a location of a single sensor, we consider the whole area affected by the seismic wave–which may include many seismic sensors–then, at distance below 1000 km, we can no longer ignore the fact that the front of the incoming wave is curved. (At larger distances from the earthquake, even at such macro-level, the curvature can be ignored.)
- It should also be mentioned that what we propose is a simple *qualitative* explanation of the observed phenomenon. To be able to explain it quantitatively–e.g., to understand why 1000 km and not any other distance is an appropriate threshold, and why exactly the Richter scale $M = 5$ is the right threshold–we probably need to supplement our simplified geometric analysis with a detailed physical analysis of the corresponding phenomena.

References

1. L. Bokati, R. Alfaro, A. Velasco, V. Kreinovich, Dynamic triggering of earthquakes: symmetry-based geometric analysis. Geombinatorics **29**(2), 78–88 (2019)
2. L. Bokati, A. Velasco, V. Kreinovich, Absence of remotely triggered large earthquakes: a geometric explanation, in *How Uncertainty-Related Ideas Can Provide Theoretical Explanation for Empirical Dependencies*. ed. by M. Ceberio, V. Kreinovich (Springer, Cham, Switzerland, 2021), pp.37–41
3. C. Aiken, Z. Peng, Dynamic triggering of microeathquakes in three geothermal/volcanic regions of California. J. Geophys. Res. **119**, 6992–7009 (2014)
4. C. Aiken, J.P. Zimmerman, J. Peng, J.I. Walter, Triggered seismic events along the Eastern Denali fault in Northwest Canada following the 2012 Mw 7.8 Haida Gwaii, 2013 Mw 7.5 Craig, and two Mw >8.5 teleseismic earthquakes. Bull. Seism. Soc. Am. **105**(2B), 1165–1177 (2015)
5. R.A. Alfaro-Diaz, *Exploring Dynamic Triggering of Earthquakes within the United States & Quarternary Faulting and Urban Seismic Hazards in the El Paso Metropolitan Area*. Ph.D. Dissertation, University of Texas at El Paso, Department of Geological Science, 2019
6. A.M. Freed, Earthquake triggering by static, dynamic, and postseismic stress transfer. Annu. Rev. Earth Planet. Sci. **33**(1), 335–367 (2005)
7. D.P. Hill, S. Prejean, Dynamic triggering, in *Earthquake Seismology*. ed. by H. Kanamori (Elsevier, Amsterdam, 2007), pp. 258–288
8. D.P. Hill, S. Prejean, Dynamic triggering, in *Earthquake Seismology*, 2nd edn., ed. by H. Kanamori (Elsevier, Amsterdam, 2015)
9. A. Kato, J. Fukuda, K. Obara, Response of seismicity to static and dynamic stress changes induced by the 2011 M9.0 Tohoku-Oki earthquake. Geophys. Res. Lett. **40**, 3572–3578 (2013)
10. K.L. Pankow, W.J. Arabasz, J.C. Pechmann, S.J. Nava, Triggered seismicity in Utah from the 3 November 2002 Denali fault earthquake. Bull. Seism. Soc. Am. **94**(6), 332–347 (2004)

11. G. Saccorotti, D. Piccinini, F. Mazzarini, M. Zupo, Remotely triggered micro-earthquakes in the Larderello-Travale geothermal field (Italy) following the 2012 May 20, Mw 5.9 Poplain earthquake. Geophys. Rev. Lett. **40**, 835–840 (2013)
12. D.R. Shelly, Z. Peng, D.P. Hill, C. Aiken, Triggered creep as a possible mechanism for delayed dynamic triggering of tremor and earthquakes. Nat. Geosci. **4**, Paper 1141 (2011)
13. A.A. Velasco, R. Alfaro-Diaz, D. Kilb, K.L. Pankow, A time-domain detection approach to identify small earthquakes within the continental United States recorded by the US Array and regional networks. Bull. Seism. Soc. Am. **106**(2), 512–525 (2016)
14. D.L. Guenaga, *Dynamic Triggering of Earthquakes Within the State of Utah, USA*, Master's Thesis, University of Texas at El Paso, Department of Geological Science, 2019
15. R. Feynman, R. Leighton, M. Sands, *The Feynman Lectures on Physics* (Addison Wesley, Boston, Massachusetts, 2005)
16. K.S. Thorne, R.D. Blandford, *Modern Classical Physics: Optics, Fluids, Plasmas, Elasticity, Relativity, and Statistical Physics* (Princeton University Press, Princeton, New Jersey, 2017)
17. E.E. Brodsky, V. Karakostas, H. Kanamori, A new observation of dynamically triggered regional seismicity: Earthquakes in Greece following the August 1999 Izmit, Turkey earthquake. Geophys. Res. Lett. **27**, 2741–2744 (2000)
18. J. Gomberg, P. Bodin, K. Larson, H. Dragert, Earthquake nucleation by transient deformations cause by the $M = 7.9$ Denali, Alaska, earthquake. Nat. **427**, 621–624 (2004)
19. J. Gomberg, P. Reasenberg, P. Bodin, R. Harris, Earthquake triggering by transient seismic waves following the Landers and Hector Mine, California, earthquake. Nat. **11**, 462–466 (2001)
20. D.P. Hill at al., Seismicity in the Western United States triggered by the M 7.4 Landers. California, earthquake of June 28, 1992. Sci. **260**, 1617–1623 (1993)
21. S.E. Hough, L. Seeber, J.G. Armbruster, Intraplate triggered earthquakes: observations and interpretation. Bull. Seism. Soc. Am. **93**(5), 2212–2221 (2003)
22. D. Kilb, J. Gomberg, P. Bodin, Triggering of earthquake aftershocks by dynamic stresses. Nat.**408**, 570–574 (2000)
23. A.A. Velasco, S. Hernandez, T. Parsons, K. Pankow, Global ubiquity of dynamic earthquake triggering. Nat. Geosci. **1**, 375–379 (2008)
24. M. West, J.J. Sanchez, S.R. McNutt, Periodically triggered seismicity at Mount Wrangell, Alaska, after the Sumatra earthquake. Sci. **308**, 1144–1146 (2005)
25. T. Parsons, A.A. Velasco, Absence of remotely triggered large earthquakes beyond the mainshock region. Nat. Geosci. **4**, 312–316 (2011)

Part IV
Applications to Teaching

In this part, we study applications to teaching. Our analysis cover all three related major questions:

- what to teach (Chaps. 20 and 21),
- how to teach (Chap. 22), and
- how to grade, i.e., how to gauge the results of teaching (Chap. 23).

Chapter 20
How Can We Explain Different Number Systems?

In this and following chapters, we study applications to teaching. In this chapter and in the next chapter, we show how general decision making techniques can help us decide on *what* to teach. In Chap. 22, we will analyze *how* to teach. And finally, in Chap. 23, we will analyze how to grade, i.e., how to gauge the results of teaching.

Specifically, in this chapter, we will analyze how to explain different number systems. At present, we mostly use decimal (base-10) number system, but in the past, many other systems were used: base-20, base-60–which is still reflected in how we divide an hour into minutes and a minute into seconds–and many others. There is a known explanation for the base-60 system: 60 is the smallest number that can be divided by 2, by 3, by 4, by 5, and by 6. Because of this, e.g., half an hour, one-third of an hour, all the way to one-sixth of an hour all correspond to a whole number of minutes. In this chapter, we show that a similar idea can explain all historical number systems, if, instead of requiring that the base divides *all* numbers from 2 to some value, we require that the base divides all but one (or all but two) of such numbers.

Comment. Results from this chapter first appeared in [1].

20.1 Formulation of the Problem

A problem. Nowadays, everyone use a decimal (base-10) system for representing integers–of course, with the exception of computers which use a binary (base-2) system. However, in the past, many cultures used different number systems; see, e.g., [2–13]. Some of these systems have been in use until reasonably recently (and are still somewhat used in colloquial speech, e.g., when we count in dozens). Other systems are only known directly from historical records or from indirect sources–such as linguistics.

L. Bokati and V. Kreinovich, *Decision Making Under Uncertainty, with a Special Emphasis on Geosciences and Education*, Studies in Systems, Decision and Control 218,
https://doi.org/10.1007/978-3-031-26086-5_20

An interesting question is: why some number systems (i.e., some bases) were used and some similar bases were not used?

Case when an explanation is known. One of the known reasons for selecting a base comes from base-60 system $\mathbf{B} = \mathbf{60}$ used by the ancient Babylonians; see, e.g., [14–16]. We still have a trace of that system–which was widely used throughout the ancient world–in our division of the hour into 60 m and a minute into 60 s.

A natural explanation for the use of this system is that it makes it easy to divide by small numbers: namely, when we divide 60 by 2, 3, 4, 5, and 6, we still get an integer. Thus, if we divide the hours into 60 m as we do, 1/2, 1/3, 1/4, 1/5, and 1/6 of the hour are all represented by a whole number of minutes–which makes it much easier for people to handle. And one can easily show that 60 is the smallest integer which is divisible by 2, 3, 4, 5, and 6.

Our idea. Let us use this explanation for the base-60 system as a sample, and see what we can get if make a similar assumption of divisibility, but for fewer numbers, or with all numbers but one or but two.

It turns out that many historically used number systems can indeed be explained this way.

20.2 Which Bases Appear If We Consider Divisibility by All Small Numbers from 1 to Some k

Let us consider which bases appear if we consider divisibility by all small natural numbers–i.e., by all natural numbers from 1 to some small number k. We will consider this for all values k from 1 to 7, and we will explain why we do not go further.

Case when $k = 2$. In this case, the smallest number divisible by 2 is the number 2 itself, so we get the binary (base-2) system $\mathbf{B} = \mathbf{2}$ used by computers.

Some cultures used powers of 2 as the base–e.g., $\mathbf{B} = \mathbf{4}$ or $\mathbf{B} = \mathbf{8}$ (see, e.g., [2]). This, in effect, is the same as using the original binary system–since, e.g., the fact that we have a special word for a hundred $100 = 10^2$ does not mean that we use a base-100 system.

Case when $k = 3$. The smallest number divisible by 2 and 3 is $\mathbf{B} = \mathbf{6}$. The base-6 number system has indeed been used, by the Morehead-Maro language of Southern New Guinea; see, e.g., [17, 18].

Case when $k = 4$. The smallest number divisible by 2, 3, and 4 is $\mathbf{B} = \mathbf{12}$. The base-12 number system has been used in many cultures; see, e.g., [7, 11, 14], and the use of dozens in many languages is an indication of this system's ubiquity.

Case when $k = 5$. The smallest number divisible by 2, 3, 4, and 5 is $B = 60$, the familiar Babylonian base. Since this number is also divisible by 6, the case $k = 6$ leads to the exact same base and thus, does not need to be considered separately.

Case when $k = 7$. The smallest number which is divisible by 2, 3, 4, 5, 6, and 7 is $B = 420$. This number looks too big to serve as the base of a number system, so we will not consider it. The same applies to larger values $k > 7$.

Thus, in this chapter, we only consider values $k \le 6$.

20.3 What If We Can Skip One Number

What happens if we consider bases which are divisible not by all, but by all-but-one numbers from 1 to k?

Of course, if we skip the number k itself, this is simply equivalent to being divisible by all the small numbers from 1 to $k - 1$–and we have already analyzed all such cases. So, it makes sense to skip a number which is smaller than k.

Let us analyze all the previous cases $k = 1, \ldots, 6$ from this viewpoint.

Case when $k = 2$. In this case, there is nothing to skip, so we still get a binary system.

Case when $k = 3$. In this case, the only number that we can skip is the number 2. The smallest integer divisible by 3 is the number 3 itself, so we get the ternary (base-3) system $\mathbf{B} = \mathbf{3}$; see, e.g., [5].

There is some evidence that people also used powers of 3, such as 9; see, e.g., [10, 19]

Case when $k = 4$. For $k = 4$, in principle, we could skip 2 or we could skip 3. Skipping 2 makes no sense, since if the base is divisible by 4, it is of course also divisible by 2 as well. Thus, the only number that we can meaningfully skip is the number 3. In this case, the smallest number which is divisible by the remaining numbers 2 and 4 is the number 4. As we have mentioned, the base-4 system is, in effect, the same as binary system–one digit of the base-4 system contains two binary digits, just like to more familiar base-8 and base-16 system, one digit corresponds to 3 or 4 binary digits.

Case when $k = 5$. In this case, we can skip 2, 3, or 4.

- Skipping 2 does not make sense, since then 4 remains, and, as we have mentioned earlier, if the base is divisible by 4, it is divisible by 2 as well.
- Skipping 3 leads to $\mathbf{B} = \mathbf{20}$, the smallest number divisible by 2, 4, and 5. Base-20 numbers have indeed been actively used, e.g., by the Mayan civilization; see, e.g., [3, 4, 14–16]. In Romance languages still 20 is described in a different way than 30, 40, and other similar numbers.
- Skipping 4 leads to $\mathbf{B} = \mathbf{30}$, the smallest number divisible by 2, 3, and 5. This seems to be the only case when the corresponding number system was not used by anyone.

Case when $k = 6$. In this case, in principle, we can skip 2, 3, 4, and 5. Skipping 2 or 3 does not make sense, since any number divisible by 6 is also divisible by 2 and 3. So, we get meaningful examples, we only consider skipping 4 or 5.

- If we skip 4, we get the same un-used base $B = 30$ that we have obtained for $k = 5$.
- If we skip 5, then the smallest number divisible by 2, 3, 4, and 6 is the base $B = 12$ which we already discussed earlier.

20.4 What If We Can Skip Two Numbers

What happens if we consider bases which are divisible by all-but-two numbers from 1 to k? Of course, to describe new bases, we need to only consider skipped numbers which are smaller than k.

Cases when $k = 2$ **or** $k = 3$. In these cases, we do not have two intermediate numbers to skip.

Case when $k = 4$. In this case, we skip both intermediate numbers 2 and 3 and consider only divisibility by 4. The smallest number divisible by 4 is the number 4 itself, and we have already considered base-4 numbers.

Case when $k = 5$. In this case, we have three intermediate numbers: 2, 3, and 4. In principle, we can form three pairs of skipped numbers: (2, 3), (2, 4), and (3, 4). Skipping the first pair makes no sense, since then 4 still remains, and if the base is divisible by 4, then it is automatically divisible by 2 as well. Thus, we have only two remaining options:

- We can skip 2 and 4. In this case, the smallest number divisible by the two remaining numbers 3 and 5 is $\mathbf{B} = \mathbf{15}$. Historically, there is no direct evidence of base-15 systems, but there is an indirect evidence: e.g., Russia used to have 15-kopeck coins, a very unusual nomination.
- We can skip 3 and 4. In this case, the smallest number divisible by the two remaining numbers 2 and 5 is $\mathbf{B} = \mathbf{10}$. This is our usual decimal system.

Case when $k = 6$. In this case, we have four intermediate values 2, 3, 4, and 5. Skipping 2 or 3 makes no sense: if the base is divisible by 6, it is automatically divisible by 2 and 3. Thus, the only pair of values that we can skip is 4 and 5. In this case, the smallest number divisible by 2, 3, and 6 is the value $B = 6$, which we have already considered earlier.

20.5 What If We Can Skip Three or More Numbers

What if we skip three numbers? What happens if we consider bases which are divisible but by all-but-three numbers from 1 to k? Of course, to describe new bases, we need to only consider skipped numbers which are smaller than k.

Cases when $k = 2, k = 3,$ **or** $k = 4$. In these cases, we do not have three intermediate numbers to skip.

Case when $k = 5$. In this case, skipping all three intermediate numbers 2, 3, and 4 leave us with $\mathbf{B} = \mathbf{5}$. The base-5 system has actually been used; see, e.g., [3].

Case when $k = 6$. In this case, we have four intermediate numbers, so skipping 3 of them means that we keep only one. It does not add to the list of bases if we keep 2 or 3, since then the smallest number divisible by 6 and by one of them is still 6–and we have already considered base-6 systems. Thus, the only options are keeping 4 and keeping 5.

If we keep 4, then the smallest number divisible by 4 and 6 is $B = 12$–our usual counting with dozens, which we have already considered.

If we keep 5, then the smallest number divisible by 5 and 6 is $B = 30$, which we have also already considered.

What if we skip more than three intermediate numbers. The only numbers $k \leq 6$ that have more than three intermediate numbers are $k = 5$ and $k = 6$.

For $k = 5$, skipping more than three intermediate numbers means skipping all fours of them, so the resulting base is $B = 5$, which we already considered.

For $k = 6$, for which there are five intermediate numbers, skipping more than three means either skipping all of them–in which case we have $B = 6$–or keeping one of the intermediate numbers. Keeping 2 or 3 still leaves us with $B = 6$, keeping 4 leads to $B = 12$, and keeping the number 5 leads to $B = 30$. All these bases have already been considered.

References

1. L. Bokati, O. Kosheleva, V. Kreinovich, How can we explain different number systems?, in *How Uncertainty-Related Ideas Can Provide Theoretical Explanation for Empirical Dependencies*. ed. by M. Ceberio, V. Kreinovich (Springer, Cham, Switzerland, 2021), pp.21–26
2. M. Ascher, *Ethnomathematics: A Multicultural View of Mathematical Ideas* (Routledge, Milton Park, Abingdon, Oxfordshire, UK, 1994)
3. T.L. Heath, *A Manual of Greek Mathematics* (Dover, New York, 2003)
4. G. Ifrah, *The Universal History of Numbers: From Prehistory to the Invention of the Computer* (John Wiley & Sons, Hoboken, New Jersey, 2000)
5. D. Knuth, *The Art of Computer Programming, Volume 2: Seminumerical Algorithms*, 3rd edn. (Addison Wesley, Boston, Massachusetts, 1998)
6. A.L. Kroeber, *Handbook of the Indians of California, Bulletin 78* (Bureau of American Ethnology of the Smithsonian Institution, Washington, DC, 1919)
7. A.R. Luria, L.S. Vygotsky, *Ape, Primitive Man, and Child: Essays in the History of Behavior* (CRC Press, Boca Raton, Florida, 1992)
8. J.P. Mallory, D.Q. Adams, *Encyclopedia of Indo-European Culture* (Fitzroy Dearborn Publsishers, Chicago and London, 1997)
9. H.J. Nissen, P. Damerow, R.K. Englund, *Archaic Bookkeeping, Early Writing, and Techniques of Economic Administration in the Ancient Near East* (University of Chicago Press, Chicago, Illinois, 1993)

10. M. Parkvall, *Limits of Language: Almost Everything You Didn't Know about Language and Languages* (William James & Company, Portland, Oregon, 2008)
11. P. Ryan, *Encyclopaedia of Papua and New Guinea* (Melbourne University Press and University of Papua and New Guinea, Melbourne, 1972)
12. D. Schmandt-Besserat, *How Writing Came About* (University of Texas Press, Austin, Texas, 1996)
13. C. Zaslavsky, *Africa Counts: Number and Pattern in African Cultures* (Chicago Review Press, Chicago, Illinois, 1999)
14. C.B. Boyer, U.C. Merzbach, *A History of Mathematics* (Wiley, New York, 1991)
15. O. Kosheleva, Mayan and Babylonian arithmetics can be explained by the need to minimize computations. Applied Mathematical Sciences **6**(15), 697–705 (2012)
16. O. Kosheleva, K. Villaverde, *How Interval and Fuzzy Techniques Can Improve Teaching* (Springer Verlag, Cham, Switzerland, 2018)
17. G. Lean, *Counting Systems of Papua New Guinea*, vols. 1–17, Papua New Guinea University of Technology, Lae, Papua New Guinea, 1988–1992
18. K. Owens, The work of glendon lean on the counting systems of papua new guinea and oceania. Math. Educ. Res. J. **13**(1), 47–71 (2001)
19. O. Kosheleva, V. Kreinovich, Was there a pre-biblical 9-ary number system? Math. Struct. Model. **50**, 87–90 (2019)

Chapter 21
Teaching Optimization

In the Appendix, we show that it is reasonable to approximate functions by polynomials. In particular, it makes sense to consider polynomial objective functions. It is therefore important to teach decision makers how to analyze such functions.

In general, people feel more comfortable with rational numbers than with irrational ones. Thus, when teaching the beginning of calculus, it is desirable to have examples of simple problems for which both zeros and extrema points are rational. Recently, an algorithm was proposed for generating cubic polynomials with this property. However, from the computational viewpoint, the existing algorithm is not the most efficient one: in addition to applying explicit formulas, it also uses trial-and-error exhaustive search. In this chapter, we describe a new computationally efficient algorithm for generating all such polynomials: namely, an algorithm that uses only explicit formulas.

Comment. Results from this chapter first appeared in [1]. The abstracts related to this result appeared in [2, 3].

21.1 Formulation of the Problem

Need for nice calculus-related examples. After students learn the basics of calculus, they practice in using the calculus tools to graph different functions $y = f(x)$. Specifically,

- they find the roots (zeros), i.e., the values where $f(x) = 0$,
- they find the extreme points, i.e., the values where the derivative is equal to 0,
- they find out whether the function is increasing or decreasing between different extreme points–by checking the signs of the corresponding derivatives,

L. Bokati and V. Kreinovich, *Decision Making Under Uncertainty, with a Special Emphasis on Geosciences and Education*, Studies in Systems, Decision and Control 218,
https://doi.org/10.1007/978-3-031-26086-5_21

and they use this information–plus the values of $f(x)$ at several points x–to graph the corresponding function.

For this practice, students need examples for which they can compute both the zeros and the extreme points.

Cubic polynomials: the simplest case when such an analysis makes sense. The simplest possible functions are polynomials. For linear functions, the derivative is constant, so there are no extreme point. For quadratic functions, there is an extreme point, but, after studying quadratic equations, students already know how to graph the corresponding function, when it decreases, when it increases. So, for quadratic polynomials, there is no need to use calculus.

The simplest case when calculus tools are needed is the case of cubic polynomials.

To make the materials simpler for students, it is desirable to limit ourselves to rational roots. Students are much more comfortable with rational numbers than with irrational ones. Thus, to make the corresponding example easier for students, it is desirable to start with examples in which all the coefficients, all the zeros, and all the extreme points of a cubic polynomial are rational.

Good news is that when we know that the roots are rational, it is (relatively) easy to find these roots. Indeed, to find rational roots, we can use the *Rational Root Theorem*, according to which for each rational root $x = p/q$ (where p and q do not have any common divisors) of a polynomial $a_n \cdot x^n + a_{n-1} \cdot x^{n-1} + \cdots + a_0$ with integer coefficients $a_0, \ldots, a_{n-1}, a_n$, the numerator p is a factor of a_0, and the denominator q is a factor of a_n; see, e.g., [4].

Thus, to find all the rational roots of a polynomial, it is sufficient:

- to list all factors p of the coefficient a_0,
- to list all factors q of the coefficient a_n, and then
- to check, for each pair (p, q) of the values from the two lists, whether the ratio p/q is a root.

How can we find polynomials for which both zeros and extreme points are rational?

What is known. An algorithm for generating such polynomials was proposed in [5, 6]. This algorithm, however, is not the most efficient one: for each tuple of the corresponding parameter values, it uses exhaustive trial-and-error search to produce the corresponding nice cubic polynomial.

What we do in this chapter. In this chapter, we produce an efficient algorithm for producing nice polynomials. Namely, we propose simple computational formulas with the following properties:

- for each tuple of the corresponding parameters, these formulas produce coefficients of a cubic polynomial for which all zeros and extreme points are rational, and
- every cubic polynomial with this property can be generated by applying these formulas to an appropriate tuple of parameters.

Thus, for each tuple of parameters, our algorithm requires the same constant number of elementary computational steps (i.e., elementary arithmetic operations)–in contrast with the existing algorithm, in which the number of elementary steps, in general, grows with the values of the parameters.

21.2 Analysis of the Problem

Let us first simplify the problem. A general cubic polynomial with rational coefficients has the form

$$a \cdot X^3 + b \cdot X^2 + c \cdot X + d. \tag{21.1}$$

We consider the case when this is a truly cubic polynomial, i.e., when $a \neq 0$.

Roots and extreme points of a function do not change if we simply divide all its values by the same constant a. Thus, instead of considering the original polynomial (21.1) with four parameters a, b, c, and d, it is sufficient to consider the following polynomial with only three parameters:

$$X^3 + p \cdot X^2 + q \cdot X + r, \tag{21.2}$$

where

$$p \stackrel{\text{def}}{=} \frac{b}{a}, \quad q \stackrel{\text{def}}{=} \frac{c}{a}, \quad r \stackrel{\text{def}}{=} \frac{d}{a}. \tag{21.3}$$

When the coefficients a, b, c, and d of the original polynomial (21.1) were rational, the coefficients of the new polynomial (21.2) are rational as well; vice versa, if we have a polynomial (21.2) with rational coefficients, then, for any rational a, we can have a polynomial (21.1) with rational coefficients $b = a \cdot p$, $c = a \cdot q$, and $d = a \cdot r$. Thus, to find cubic polynomials with rational coefficients, rational roots, and rational extreme points, it is sufficient to consider polynomials of type (21.2).

We can simplify the problem even further if we replace the original variable X with the new variable

$$x \stackrel{\text{def}}{=} X + \frac{p}{3} \tag{21.4}$$

for which

$$X = x - \frac{p}{3}. \tag{21.5}$$

Substituting this expression for X into the formula (21.2), we get

$$\left(x-\frac{p}{3}\right)^3+p\cdot\left(x-\frac{p}{3}\right)^2+q\cdot\left(x-\frac{p}{3}\right)+r=$$

$$x^3-3\cdot\frac{p}{3}\cdot x^2+3\cdot\left(\frac{p}{3}\right)^2\cdot x-\left(\frac{p}{3}\right)^3+p\cdot x^2-$$

$$2\cdot p\cdot\frac{p}{3}\cdot x+p\cdot\left(\frac{p}{3}\right)^2+q\cdot x-q\cdot\frac{p}{3}+r=$$

$$x^3+\alpha\cdot x+\beta, \tag{21.6}$$

where

$$\alpha=q-\frac{p^2}{3} \tag{21.7}$$

and

$$\beta=r-\frac{p\cdot q}{3}+\frac{2p^3}{27}. \tag{21.8}$$

The roots and extreme points of the new polynomial (21.6) are obtained from the roots and extremes of the original polynomial (21.2) by shifting by a rational number $p/3$, so they are all rational for the polynomial (21.6) if and only if they are rational for the polynomial (21.2).

Describing in terms of roots. Let r_1, r_2, and r_3 denote rational roots of the polynomial (21.6). Then, we have

$$x^3+\alpha\cdot x+\beta=(x-r_1)\cdot(x-r_2)\cdot(x-r_3)=$$

$$x^3-(r_1+r_2+r_3)\cdot x^2+(r_1\cdot r_2+r_2\cdot r_3+r_1\cdot r_3)\cdot x-r_1\cdot r_2\cdot r_3. \tag{21.9}$$

By equating the coefficients at x^2, x, and 1 at both sides, we conclude that

$$r_1+r_2+r_3=0, \tag{21.10}$$

$$\alpha=r_1\cdot r_2+r_2\cdot r_3+r_1\cdot r_3, \tag{21.11}$$

and

$$\beta=-r_1\cdot r_2\cdot r_3. \tag{21.13}$$

From (21.10), we conclude that

$$r_3=-(r_1+r_2). \tag{21.14}$$

Substituting the expression (21.14) into the formulas (21.11) and (21.13), we conclude that

$$\alpha = r_1 \cdot r_2 - r_2 \cdot (r_1 + r_2) - r_1 \cdot (r_1 + r_2) = -(r_1^2 + r_1 \cdot r_2 + r_2^2) \tag{21.15}$$

and

$$\beta = r_1 \cdot r_2 \cdot (r_1 + r_2). \tag{21.16}$$

Now the polynomial (21.6) takes the following form:

$$x^3 - (r_1^2 + r_1 \cdot r_2 + r_2^2) \cdot x + r_1 \cdot r_2 \cdot (r_1 + r_2). \tag{21.17}$$

Using the fact that the extreme points should also be rational. Let us now use the fact that the extreme points should also be rational. Let x_0 denote an extreme point, i.e., a point at which the derivative of the polynomial (21.17) is equal to 0. Differentiating the expression (21.17) and equating the derivative to 0, we get

$$3x_0^2 - (r_1^2 + r_1 \cdot r_2 + r_2^2) = 0. \tag{21.18}$$

The expression in parentheses can be equivalently described as

$$\frac{3}{4} \cdot (r_1 + r_2)^2 + \frac{1}{4} \cdot (r_1 - r_2)^2 = 3y^2 + z^2, \tag{21.19}$$

where we denoted

$$y \stackrel{\text{def}}{=} \frac{r_1 + r_2}{2} \text{ and } z \stackrel{\text{def}}{=} \frac{r_1 - r_2}{2}. \tag{21.20}$$

Substituting this expression (21.20) into the formula (21.18), we arrive at the following homogeneous quadratic relation with integer coefficients between the rational numbers x_0, y, and z:

$$3x_0^2 - 3y^2 - z^2 = 0. \tag{21.21}$$

If we divide both sides of Eq. (21.21) by y^2, we get a new equation

$$3X_0^2 - 3 - Z^2 = 0, \tag{21.22}$$

where we denoted $X_0 \stackrel{\text{def}}{=} \frac{x_0}{y}$ and $Z \stackrel{\text{def}}{=} \frac{z}{y}$. When x_0, y, and z are rational, then X_0 and Z are also rational numbers. Vice versa, when X_0 and Z form a rational-valued solution of the Eq. (21.22), then, for any rational number y, by multiplying both sides of Eq. (21.22) by y^2, we can get a solution $x_0 = y \cdot X_0$, y, and $z = y \cdot Z$ of the Eq. (21.21). Thus, to find all rational solutions of the Eq. (21.21), it is sufficient to find all rational solutions of a simplified Eq. (21.22).

The simplest solution and the resulting "nice" polynomials. One of the solution of Eq. (21.22) is easy to find: namely, when $X_0 = -1$, the Eq. (21.22) takes the form $Z^2 = 0$, i.e., $Z = 0$.

This means that for every y, the values $x_0 = -y$, y and $z = 0$ solve the Eq. (21.21). The formulas (21.20) enable us to reconstruct r_1 and r_2 from y and z as

$$r_1 = y + z \text{ and } r_2 = y - z. \tag{21.23}$$

In our case, this means $r_1 = r_2 = y$. Thus, due to (21.15) and (21.16), we have a polynomial $x^3 + \alpha \cdot x + \beta$ with $\alpha = -3y^2$ and $\beta = 2y^3$.

By applying a shift by a rational number s, i.e., by replacing x with $x = X + s$, we transform a "nice" polynomial $x^3 + \alpha \cdot x + \beta$ into a new "nice" polynomial

$$(X + s)^3 + \alpha \cdot (X + s) + \beta = X^3 + 3s \cdot X^2 + (3s^2 + \alpha) \cdot X + (s^3 + \beta + \alpha \cdot s),$$

i.e., a polynomial (21.2) with $p = 3s$, $q = 3s^2 + \alpha$, and $r = s^3 + \beta$. Finally, by multiplying this polynomial by a rational number a, we get the following family of "nice" polynomials:

$$b = 3a \cdot s, \quad c = a \cdot (3s^2 + \alpha), \quad d = a \cdot (s^3 + \beta + \alpha \cdot s). \tag{21.24}$$

In our case, with $\alpha = -3y^2$ and $\beta = 2y^3$, we get

$$b = 3a \cdot s, \quad c = a \cdot (3s^2 - 3y^2), \quad d = a \cdot (s^3 + 2y^3 - 3y^2 \cdot s). \tag{21.24a}$$

Using the general algorithm for finding all rational solutions to a quadratic equation. To find all rational solutions of the Eq. (21.21), we will use a general algorithm for finding all rational solutions of a homogeneous quadratic equation with integer coefficients; see, e.g., [7].

We have already found a solution of the Eq. (21.22) corresponding to $X_0 = -1$. For this value X_0, the Eq. (21.22) has only one solution $(-1, 0)$, for which $X_0 = -1$ and $Z = 0$. Every other solution (X_0, Z) can be connected to this simple solution $(-1, 0)$ by a straight line. A general equation of a straight line passing through the point $(-1, 0)$ is

$$Z = t \cdot (X_0 + 1). \tag{21.25}$$

When X_0 and Z are rational, the ratio $t = \dfrac{Z}{X_0 + 1}$ is also rational.

Substituting the expression (21.25) into the Eq. (21.22), we get

$$3X_0^2 - 3 - t^2 \cdot (X_0 + 1)^2 = 0,$$

i.e.,

$$3 \cdot (X_0^2 - 1) - t^2 \cdot (X_0 + 1)^2 = 0. \tag{21.26}$$

Since we consider the case when $X_0 \neq -1$, we thus have $X_0 + 1 \neq 0$. So, we can divide both sides of the Eq. (21.26) by $X_0 + 1$ and thus, get the following equation:

$$3 \cdot (X_0 - 1) - t^2 \cdot (X_0 + 1) = 0.$$

From this equation, we can describe X_0 in terms of t: $(3 - t^2) \cdot X_0 = 3 + t^2$, hence

$$X_0 = \frac{3 + t^2}{3 - t^2}. \tag{21.27}$$

Substituting this expression for X_0 into the formula (21.25), we conclude that

$$Z = \frac{6t}{3 - t^2}. \tag{21.28}$$

Towards a general description of all "nice" polynomials. For every rational y, we can now take $x_0 = y \cdot X_0$, y, and

$$z = y \cdot Z = \frac{6yt}{3 - t^2}. \tag{21.29}$$

Based on y and z, we can compute r_1 and r_2 by using the formulas (21.23).

We can now use the values r_1 and r_2 from (21.23) and the formulas (21.15) and (21.16) to compute α and β. Since here, $r_1 + r_2 = 2y$, we get

$$\alpha = r_1 \cdot r_2 - (r_1 + r_2)^2 = (y + z) \cdot (y - z) - (2y)^2 =$$

$$y^2 - z^2 - 4y^2 = -3y^2 - z^2 \tag{21.30}$$

and

$$\beta = r_1 \cdot r_2 \cdot (r_1 + r_2) = (y^2 - z^2) \cdot (2y) = 2y \cdot (y^2 - z^2). \tag{21.31}$$

Substituting these expressions for α and β into the formula (21.24), we get the formulas for computing the coefficients of the corresponding "nice" cubic polynomial:

$$b = 3a \cdot s; \tag{21.32}$$

$$c = a \cdot (3s^2 + \alpha) = a \cdot (3s^2 - 3y^2 - z^2); \tag{21.33}$$

$$d = a \cdot (s^3 + \beta + \alpha \cdot s) = a \cdot (s^3 + 2y \cdot (y^2 - z^2) - (3y^2 + z^2) \cdot s). \tag{21.34}$$

Thus, we arrive at the following algorithm for computing all possible "nice" cubic polynomials.

21.3 Resulting Algorithm

Here is an algorithm for computing all "nice" cubic polynomials, i.e., all cubic polynomials with rational coefficients for which all three roots and both extreme points are rational.

In this algorithm, we use four arbitrary rational numbers t, y, s, and a. Based on these numbers, we first compute

$$z = \frac{6yt}{3 - t^2}. \tag{21.29a}$$

Then, we compute the coefficients b, c, and d of the resulting "nice" polynomial (the value a we already know):

$$b = 3a \cdot s; \tag{21.32}$$

$$c = a \cdot (3s^2 - 3y^2 - z^2); \tag{21.33a}$$

$$d = a \cdot (s^3 + 2y \cdot (y^2 - z^2) - (3y^2 + z^2) \cdot s). \tag{21.34a}$$

These expressions cover almost all "nice" polynomials, with the exception of one family of such polynomials, which is described by the formula

$$b = 3a \cdot s, \quad c = a \cdot (3s^2 - 3y^2), \quad d = a \cdot (s^3 + 2y^3 - 3y^2 \cdot s). \tag{21.24a}$$

References

1. L. Bokati, O. Kosheleva, V. Kreinovich, How to generate 'nice' cubic polynomials - with rational coefficients, rational zeros and rational extrema: a fast algorithm. J. Uncertain Syst. **13**(2), 94–99 (2019)
2. L. Bokati, O. Kosheleva, V. Kreinovich, How to generate 'nice' cubic polynomials – with rational coefficients, rational zeros and rational extrema: a fast algorithm, in *Abstracts of the 11th International Conference of Nepalese Student Association NeSA'11*, Las Cruces, New Mexico, 23 Mar 2019
3. L. Bokati, O. Kosheleva, V. Kreinovich, How to generate 'nice' cubic polynomials – with rational coefficients, rational zeros and rational extrema: a fast algorithm, in *Abstracts of the NMSU/UTEP Workshop on Mathematics, Computer Science, and Computational Science*, Las Cruces, New Mexico, 6 Apr 2019
4. C.D. Miller, M.L. Lial, D.I. Schneider, *Fundamentals of College Algebra*, Scott and Foresman/Little and Brown Higher Education, 1990
5. C.L. Adams, Introducing roots and extrema in calculus: use cubic polynomial functions before increasing the difficulty with irrational values. Math. Teach. **112**(2), 132–135 (2018)

6. C.L. Adams, J. Board, Conditions on a coefficients of a reduced cubic polynomial such that it and its derivative are factorable over the rational numbers, in *Electronic Proceedings of the 28th Annual International Conference on Technology in Collegiate Mathematics*, Atlanta, Georgia, 10–13 Mar 2016, pp. 33–45. http://archives.mat.utk.edu/ICTCM/i/28/A003.html
7. N.P. Smart, *The Algorithmic Resolution of Diophantine Equations* (Cambridge University Press, Cambridge, UK, 1998)

Chapter 22
Why Immediate Repetition is Good for Short-Time Learning Results But Bad for Long-Time Learning: Explanation Based on Decision Theory

In this chapter, we apply decision making under uncertainty to decide how to teach. Specifically, it is well known that repetition enhances learning; the question is: when is a good time for this repetition? Several experiments have shown that immediate repetition of the topic leads to better performance on the resulting test than a repetition after some time. Recent experiments showed, however, that while immediate repetition leads to better results on the test, it leads to much worse performance in the long term, i.e., several years after the material have been studied. In this chapter, we use decision theory to provide a possible explanation for this unexpected phenomenon.

Comment. Results from this chapter first appeared in [1].

22.1 Formulation of the Problem: How to Explain Recent Observations Comparing Long-Term Results of Immediate and Delayed Repetition

Repetitions are important for learning. A natural idea to make students better understand and better learn the material is to repeat this material–the more times we repeat, the better the learning results.

This repetition can be explicit–e.g., when we go over the material once again before the test. This repetition can be implicit–e.g., when we give the students a scheduled quiz on the topic, so that they repeat the material themselves when preparing for this quiz.

When should we repeat? The number of repetitions is limited by the available time. Once the number of repetitions is fixed, it is necessary to decide when should we have a repetition:

L. Bokati and V. Kreinovich, *Decision Making Under Uncertainty, with a Special Emphasis on Geosciences and Education*, Studies in Systems, Decision and Control 218,
https://doi.org/10.1007/978-3-031-26086-5_22

- shall we have it immediately after the students have studied the material, or
- shall we have it after some time after this studying, i.e., after we have studied something else.

What was the recommendation until recently. Experiments have shown that repeating the material almost immediately after the corresponding topic was first studied–e.g., by giving a quiz on this topic–enhances the knowledge of this topic that the students have after the class as a whole. This enhancement was much larger than when a similar quiz–reinforcing the students' knowledge–was given after a certain period of time after studying the topic.

New data seems to promote the opposite recommendation. This idea has been successfully used by many instructors. However, a recent series of experiments has made many researchers doubting this widely spread strategy. Specifically, these experiments show that (see, e.g., [2] and references therein):

- while immediate repetition indeed enhances the amount of short-term (e.g., semester-wide) learning more than a later repetition,
- from the viewpoint of long-term learning–what the student will be able to recall in a few years (when he or she will start using this knowledge to solve real-life problems)–the result is opposite: delayed repetitions lead to much better long-term learning than the currently-fashionable immediate ones.

Why? The above empirical result is somewhat unexpected, so how can we explain it? We have partially explained the advantages of *interleaving*–a time interval between the study and the repetition–from the general geometric approach; see, e.g., [3, 4]. However, this explanation does not cover the difference between short-term and long-term memories.

So how can we explain this observed phenomenon? We can simply follow the newer recommendations, kind of arguing that human psychology is difficult, has many weird features, so we should trust whatever the specialists tell us. This may sound reasonable at first glance, but the fact that we have followed this path in the past and came up with what seems now to be wrong recommendation–this fact encourages us to take a pause, and first try to understand the observed phenomenon, and only follow it if it makes sense.

This is exactly the purpose of this chapter: to provide a reasonable explanation for the observed phenomenon.

22.2 Main Idea Behind Our Explanation: Using Decision Theory

Main idea: using decision theory. Our memory is limited in size. We cannot memorize everything that is happening to us. Thus, our brain needs to decide what to store in a short-term memory, what to store in a long-term memory, and what not to store at all.

How can we make this decision? There is a whole area of research called *decision theory* that describes how we make decisions–or, to be more precise, how a rational person should make decisions.

Usually, this theory is applied to conscientious decisions, i.e., decisions that we make after some deliberations. However, it is reasonable to apply it also to decisions that we make on subconscious level–e.g., to decisions on what to remember and what not to remember: indeed, these decisions should also be made rationally.

Let us apply this to learning. If we learn the material, we spend some resources on storing it in memory. If we do not learn the material, we may lose some utility next time when this material will be needed. So, whether we store the material in memory depends on for which of the two possible actions–to learn or not to learn–utility is larger (or equivalently, losses are smaller). Let us describe this idea in detail.

22.3 So When Do We Learn: Analysis of the Problem and the Resulting Explanation

Notations. To formalize the above idea, let us introduce some notations.

- Let m denote the losses (= negative utility) needed to store a piece of material in the corresponding memory (short-term or long-term).
- Let L denote losses that occur when we need this material but do not have it in our memory.
- Finally, let p denote our estimate of the probability that this material will be needed in the corresponding time interval (short-term time interval for short-term memory or long-term time interval for long-term memory).

If we learn, we have loss m. If we do not learn, then the expected loss is equal to $p \cdot L$. We learn the material if the second loss of larger, i.e., if $p \cdot L > m$, i.e., equivalently, if $p > m/L$.

Comment. Sometimes, students underestimate the usefulness of the studied material, i.e., underestimate the value L. In this case, L is low, so the ratio m/L is high, and for most probability estimates p, learning does not make sense. This unfortunate situation can be easily repaired if we explain, to the students, how important this knowledge can be–and thus, make sure that they estimate the potential loss L correctly.

Discussion. For different pieces of the studied material, we have different ratios m/L. These ratios do not depend on the learning technique. As we will show later, the estimated probability p may differ for different learning techniques. So, if one technique consistently leads to higher values p, this means that, in general, that for more pieces of material we will have $p > m/L$ and thus, more pieces of material will be learned. So, to compare two different learning techniques, we need to compare the corresponding probability estimates p.

Let us formulate the problem of estimating the corresponding probability p in precise terms.

Towards a precise formulation of the probability estimation problem. In the absence of other information, to estimate the probability that this material will be needed in the future, the only information that our brain can use is that there were two moments of time at which we needed this material in the past:

- the moment t_1 when the material was first studied, and
- the moment t_2 when the material was repeated.

In the immediate repetition case, the moment t_2 was close to t_1, so the difference $t_2 - t_1$ was small. In the delayed repetition case, the difference $t_2 - t_1$ is larger.

Based on this information, the brain has to estimate the probability that there will be another moment of time during some future time interval. How can we do that?

Let us first consider a deterministic version of this problem. Before we start solving the actual probability-related problem, let us consider the following simplified deterministic version of this problem:

- we know the times $t_1 < t_2$ when the material was needed;
- we need to predict the next time t_3 when the material will be needed.

We can reformulate this problem in more general terms:

- we observed some event at moments t_1 and $t_2 > t_1$;
- based on this information, we want to predict the moment t_3 at which the same event will be observed again.

In other words, we need to have a function $t_3 = F(t_1, t_2) > t_2$ that produces the desired estimate.

What are the reasonable properties of this prediction function? The numerical value of the moment of time depends on what unit we use to measure time–e.g., hours, days, or months. It also depends on what starting point we choose for measuring time. We can measure it from Year 0 or–following Muslim or Buddhist calendars–from some other date.

If we replace the original measuring unit with the new one which is a times smaller, then all numerical values will multiply by a: $t \to t' = a \cdot t$. For example, if we replace seconds with milliseconds, all numerical values will multiply by 1000, so, e.g., 2 sec will become 2000 msec. Similarly, if we replace the original starting point with the new one which is b units earlier, then the value b will be added to all numerical values: $t \to t' = t + b$. It is reasonable to require that the resulting prediction t_3 not depend on the choice of the unit and on the choice of the starting point. Thus, we arrive at the following definitions.

Definition 22.1 We say that a function $F(t_1, t_2)$ is *scale-invariant* if for every t_1, t_2, t_3, and $a > 0$, if $t_3 = F(t_1, t_2)$, then for $t'_i = a \cdot t_i$, we get $t'_2 = F(t'_1, t'_2)$.

Definition 22.2 We say that a function $F(t_1, t_2)$ is *shift-invariant* if for every t_1, t_2, t_3, and b, if $t_3 = F(t_1, t_2)$, then for $t_i' = t_i + b$, we get $t_3' = F(t_1', t_2')$.

Proposition 22.1 *A function $F(t_1, t_2) > t_2$ is scale-and shift-invariant if and only if it has the form $F(t_1, t_2) = t_2 + \alpha \cdot (t_2 - t_1)$ for some $\alpha > 0$.*

Proof Let us denote $\alpha \stackrel{\text{def}}{=} F(-1, 0)$. Since $F(t_1, t_2) > t_2$, we have $\alpha > 0$. Let $t_1 < t_2$, then, due to scale-invariance with $a = t_2 - t_1 > 0$, the equality $F(-1, 0) = \alpha$ implies that $F(t_1 - t_2, 0) = \alpha \cdot (t_2 - t_1)$. Now, shift-invariance with $b = t_2$ implies that $F(t_1, t_2) = t_2 + \alpha \cdot (t_2 - t_1)$. The proposition is proven. □

Discussion. Many physical processes are reversible: if we have a sequence of three events occurring at moments $t_1 < t_2 < t_3$, then we can also have a sequence of events at times $-t_3 < -t_2 < -t_1$. It is therefore reasonable to require that:

- if our prediction works for the first sequence, i.e., if, based on t_1 and t_2, we predict t_3,
- then our prediction should work for the second sequence as well, i.e. based on $-t_3$ and $-t_2$, we should predict the moment $-t_1$.

Let us describe this requirement in precise terms.

Definition 22.3 We say that a function $F(t_1, t_2)$ is *reversible* if for every t_1, t_2. and t_3, the equality $F(t_1, t_2) = t_3$ implies that $F(-t_3, -t_2) = -t_1$.

Proposition 22.2 *The only scale-and shift-invariant reversible function $F(t_1, t_2)$ is the function $F(t_1, t_2) = t_2 + (t_2 - t_1)$.*

Comment. In other words, if we encounter two events separated by the time interval $t_2 - t_1$, then the natural prediction is that the next such event will happen after exactly the same time interval.

Proof In view of Proposition 22.1, all we need to do is to show that for a reversible function we have $\alpha = 1$. Indeed, for $t_1 = -1$ and $t_2 = 0$, we get $t_3 = \alpha$. Then, due to Proposition 22.1, we have $F(-t_3, -t_2) = F(-\alpha, 0) = 0 + \alpha \cdot (0 - (-\alpha)) = \alpha^2$. The requirement that this value should be equal to $-t_1 = 1$ implies that $\alpha^2 = 1$, i.e., due to the fact that $\alpha > 0$, that $\alpha = 1$. The proposition is proven. □

From simplified deterministic case to the desired probabilistic case. In practice, we cannot predict the actual time t_3 of the next occurrence, we can only predict the *probability* of different times t_3. Usually, the corresponding uncertainty is caused by a joint effect of many different independent factors. It is known that in such situations, the resulting probability distribution is close to Gaussian–this is the essence of the Central Limit Theorem which explains the ubiquity of Gaussian distributions; see, e.g., [5]. It is therefore reasonable to conclude that the distribution for t_3 is Gaussian, with some mean μ and standard deviation σ.

There is a minor problem with this conclusion; namely:

- Gaussian distribution has non-zero probability density for all possible real values, while
- we want to have only values $t_3 > t_2$.

This can be taken into account if we recall that in practice, values outside a certain $k\sigma$-interval $[\mu - k \cdot \sigma, \mu + k \cdot \sigma]$ have so little probability that they are considered to be impossible. Depending on how low we want this probability to be, we can take $k = 3$, or $k = 6$, or some other value k. So, it is reasonable to assume that the lower endpoint of this interval corresponds to t_2, i.e., that $\mu - k \cdot \sigma = t_2$. Hence, for given t_1 and t_2, once we know μ, we can determine σ. Thus, to find the corresponding distribution, it is sufficient to find the corresponding value μ.

As this mean value μ, it is reasonable to take the result of the deterministic prediction, i.e., $\mu = t_2 + (t_2 - t_1)$. In this case, from the above formula relating μ and σ, we conclude that $\sigma = (t_2 - t_1)/k$.

Finally, an explanation. Now we are ready to explain the observed phenomenon.

In the case of immediate repetition, when the difference $t_2 - t_1$ is small, most of the probability–close to 1–is located is the small vicinity of t_1, namely in the $k\sigma$ interval which now takes the form $[t_2, t_2 + 2(t_2 - t_1)]$. Thus, in this case, we have:

- (almost highest possible) probability $p \approx 1$ that the next occurrence will happen in the short-term time interval and
- close to 0 probability that it will happen in the long-term time interval.

Not surprisingly, in this case, we get:

- a better short-term learning than for other learning strategies, but
- we get much worse long-term learning.

In contrast, in the case of delayed repetition, when the difference $t_2 - t_1$ is large, the interval $[t_2, t + 2(t_2 - t_1)]$ of possible values t_3 spreads over long-term times as well. Thus, here:

- the probability p to be in the short-time interval is smaller than the value ≈ 1 corresponding to immediate repetition, but
- the probability to be in the long-term interval is larger that the value ≈ 0 corresponding to immediate repetition.

As a result, for this learning strategy:

- we get worse short-term learning but
- we get much better long-term learning,

exactly as empirically observed.

References

1. L. Bokati, J. Urenda, O. Kosheleva, V. Kreinovich, Why immediate repetition is good for short-term learning results but bad for long-term learning: explanation based on decision theory, in *How Uncertainty-Related Ideas Can Provide Theoretical Explanation for Empirical Dependencies*. ed. by M. Ceberio, V. Kreinovich (Springer, Cham, Switzerland, 2021), pp.27–35
2. D. Epstein, *Range: Why Generalists Triumph in the Specialized World* (Riverhead Books, New York, 2019)
3. O. Kosheleva, K. Villaverde, *How Interval and Fuzzy Techniques Can Improve Teaching* (Springer Verlag, Cham, Switzerland, 2018)
4. O. Lerma, O. Kosheleva, V. Kreinovich, Interleaving enhances learning: a possible geometric explanation. Geombinatorics. **24**(3), 135–139 (2015)
5. D.J. Sheskin, *Handbook of Parametric and Non-Parametric Statistical Procedures* (Chapman & Hall/CRC, London, UK, 2011)

Chapter 23
How to Assign Grades to Tasks so as to Maximize Student Efforts

In this chapter, we use decision making under uncertainty to decide which grading system leads to the most effective teaching. Specifically, in some classes, students want to get a desired passing grade (e.g., C or B) by spending the smallest amount of effort. In such situations, it is reasonable for the instructor to assign the grades for different tasks in such a way that the resulting overall student's effort is the largest possible. In this chapter, we show that to achieve this goal, we need to assign, to each task, the number of points proportional to the efforts needed for this task.

Comment. Results from this chapter first appeared in [1].

23.1 Formulation of the Problem

In some cases, students try to minimize their efforts. In the ideal world, students should apply the maximal effort when studying for all their classes. In reality, students usually have a limited amount of time. As a result, while they concentrate their efforts on their major classes, they limit their efforts in other classes to a necessary minimum–usually, the minimum effort needed to get a passing grade in this class.

This phenomenon is especially frequent when students take required classes outside their major discipline–e.g., when engineering students take required humanity classes or when biology majors take math and/or computing classes which are not directly related to their discipline.

How can instructors increase the students' efforts in these classes. For classes in which students minimize their efforts, instructors try to maximize the student efforts–to make sure that even with the current attitude, the students learn as much

L. Bokati and V. Kreinovich, *Decision Making Under Uncertainty, with a Special Emphasis on Geosciences and Education*, Studies in Systems, Decision and Control 218,
https://doi.org/10.1007/978-3-031-26086-5_23

of the topic as possible. Since all these students care about is their overall grade for this class, the only thing that the instructor controls is which proportion of the grade goes for each task. How can we assign these grades so as to maximize the student efforts?

Towards a precise formulation of the problem. The overall grade for the classes is usually computed as a weighted average of grades for different tasks, i.e., in effect, as the sum of partial grades given for each task. The ideal case–usually described by $I = 100$ points–corresponds to the case when the student gets the maximum possible number of points for each of the tasks.

Let n denote the total number of tasks, and let m_i denote the maximum number of points that a student can get for each task. Then, we have $I = \sum_{i=1}^{n} m_i$. Let e_i denote the amount of effort (e.g., measured by the time of intensive study) that a student needs to get the maximum number of point m_i in the i-th task, and let $E = \sum_{i=1}^{n} e_i$ denote the overall effort needed to get a perfect grade m_i on all the tasks–and thus, the perfect grade I for the class.

As we have mentioned, the students do not always apply the maximum effort in studying. Let a_i be the actual effort that the student applies to the i-th task–e.g., into studying the i-th part of the material. (A student may be studying more than needed, but we only count the time that the student studies for the corresponding task.) Since the effort e_i already provides a perfect mastery of the i-th task, we assume that $a_i \le e_i$.

In the first approximation, it is reasonable to assume that the number of points gained by the student is proportional to the student's effort. If the student applies the maximal effort e_i, this student will get m_i points. Thus, in general, for each effort a_i, the resulting number of points g_i is equal to $g_i = a_i \cdot \frac{m_i}{e_i}$. The student wants to minimize the overall effort $\sum_{i=1}^{n} a_i$ under the constraints that the overall number of points is greater than or equal to the passing value g_0: $\sum_{i=1}^{n} a_i \cdot \frac{m_i}{e_i} \ge g_0$. Thus, we arrive at the following precise formalization of the problem.

Precise formulation of the problem. Let us assume that we are given values I, $e_1, \ldots, e_n$, and g_0. For each tuple $m = (m_1, \ldots, m_n)$ for which $\sum_{i=1}^{n} m_i = I$, let $E(m)$ denote the value $\sum_{i=1}^{n} a_i$ corresponding to the solution to the following constraint optimization problem:

$$\sum_{i=1}^{n} a_i \to \min_{a_1, \ldots, a_n} \tag{23.1}$$

under the constraints

$$0 \le a_i \le e_i \tag{23.2}$$

and

$$\sum_{i=1}^{n} a_i \cdot \frac{m_i}{e_i} \geq g_0. \tag{23.3}$$

Our goal is to select a tuple m for which the corresponding overall effort $E(m)$ is the largest possible:

$$E(m) \to \max_{m}. \tag{23.4}$$

Comment. This problem is an example of a *bilevel* optimization problem, in which on the top level, we select the parameters of the objective functions so that the solution to the resulting *low-level* optimization problem will optimize an appropriate high-level objective function; see, e.g., [2].

In our case, the low level optimization is performed by a student, who is trying to minimize his/her efforts under the constraint that his overall number of points is at least g_0. The grades m_i for each task are parameters in this student's optimization problem. The instructor–top-level optimizer–would like to select these parameters in such a way that the resulting overall student's effort is as large as possible.

23.2 Solution to the Problem

Description of the solution. As we show in this section, the optimal solution is to assign grades m_i proportional to the effort, i.e., to have

$$m_i = \frac{I}{E} \cdot e_i. \tag{23.5}$$

Proof For the above assignment, we have $\frac{m_i}{e_i} = \frac{I}{E}$, so for all possible actual efforts a_i, the resulting grade is equal to $\sum_{i=1}^{n} \frac{m_i}{e_i} \cdot a_i = \frac{I}{E} \cdot \sum_{i=1}^{n} a_i$ and is, thus, proportional to the overall effort $A = \sum_{i=1}^{n} a_i$. The student wants to minimize the overall effort under the condition that this grade is at least g_0. The corresponding constraint $\frac{I}{E} \cdot A \geq g_0$ is equivalent to $A \geq g_0 \cdot \frac{E}{I}$. Thus, the smallest possible value $E(m)$ of the overall effort A is equal to

$$E(m) = g_0 \cdot \frac{E}{I}. \tag{23.6}$$

Let us prove that for any other grade assignment $m' \neq m$, we have

$$E(m') < E(m) = g_0 \cdot \frac{E}{I}.$$

Indeed, the assignment m is characterized by the fact that for this assignment, the ratio $\frac{m_i}{e_i}$ is constant. Since $m' \neq m$, for this new assignment, the ratio $\frac{m'_i}{e_i}$ is not constant, it takes at least two different values for some i.

If we had $\frac{m'_i}{e_i} \leq \frac{I}{E}$ for all i, then, since $m' \neq m$, we should have $\frac{m'_j}{e_j} < \frac{I}{E}$ for some j. In this case, we have $m'_i \leq e_i \cdot \frac{I}{E}$ for all i and $m'_j < e_j \cdot \frac{I}{E}$ for some j. By adding all these inequalities, we get $\sum_{i=1}^{n} m'_i < E \cdot \frac{I}{E} = I$, which contradicts the fact that for each grade assignment, we should have $\sum_{i=1}^{n} m'_i = I$. Thus, this case is impossible, and we have at least one index i for which $\frac{m'_i}{e_i} > \frac{I}{E}$. Let us denote one of these indices by k, then $\frac{m'_k}{e_k} > \frac{I}{E}$ and $m'_k > e_k \cdot \frac{I}{E}$. If we subtract this inequality from the equality $I = E \cdot \frac{I}{E}$, then we get $I - m'_k < (E - e_k) \cdot \frac{I}{E}$, hence $\frac{I - m'_k}{E - e_k} < \frac{I}{E}$. From this inequality and the inequality $\frac{m'_k}{e_k} < \frac{I}{E}$, we conclude that $\frac{I - m'_k}{E - e_k} < \frac{m'_k}{e_k}$. Taking the inverse of both sides, we conclude that

$$\frac{e_k}{m'_k} < \frac{E - e_k}{I - m'_k},$$

thus

$$e_k < m'_k \cdot \frac{E - e_k}{I - m'_k}. \tag{23.7}$$

For each $\varepsilon > 0$ and $\delta > 0$, let the student spend a little bit more effort on the k-th assignment than in the proportional assignment, i.e., $a_k = \left(\frac{g_0}{I} + \varepsilon\right) \cdot e_k$, while for all other tasks $i \neq k$, the student will spend a little less effort $a_i = \left(\frac{g_0}{I} - \delta\right) \cdot e_i$. Under the grade assignment m', the student's grade g will be equal to

$$g = \left(\frac{g_0}{I} + \varepsilon\right) \cdot m'_k + \sum_{i \neq k} \left(\frac{g_0}{I} - \delta\right) \cdot m'_i.$$

Opening parentheses, combining terms proportional to $\frac{g_0}{I}$, and taking into account that $m_k + \sum_{i \neq k} m'_i = I$ and thus $\sum_{i \neq k} m'_i = I - m'_k$, we conclude that

$$g = \frac{g_0}{I} \cdot I + \varepsilon \cdot m'_k - \delta \cdot (M - m'_k) = g_0 + \varepsilon \cdot m'_k - \delta \cdot (M - m'_k).$$

We can get $g = g_0$ if we select δ in such a way that $\varepsilon \cdot m'_k - \delta \cdot (M - m'_k) = 0$. Then,

$$\delta = \varepsilon \cdot \frac{m'_k}{I - m'_k}. \tag{23.8}$$

For this selection of δ, the student's overall effort is equal to

$$A = \sum_{i=1}^{n} a_i = a_k + \sum_{i \neq k} a_i = \left(\frac{g_0}{I} + \varepsilon\right) \cdot e_k + \sum_{i \neq k} \left(\frac{g_0}{I} - \delta\right) \cdot e_i.$$

Opening the parentheses, combining terms proportional to $\frac{g_0}{I}$, and taking into account that $e_k + \sum_{i \neq k} e_i = E$ and thus $\sum_{i \neq k} e_i = E - e_k$, we conclude that

$$A = g_0 \cdot \frac{E}{I} + \varepsilon \cdot e_k - \delta \cdot (E - e_k).$$

According to the formula (23.4), the first term in the right-hand side is exactly $E(m)$ for the above grade assignment m. Substituting the expression (23.8) for δ into this formula, we conclude that

$$A = E(m) + \varepsilon \cdot \left(e_k - \frac{m'_k}{I - m'_k} \cdot (E - e_k)\right).$$

Due to (23.7), we have $A < E(m)$. By definition, $E(m')$ is the smallest possible effort the student needs to spend to get g_0, thus $E(m') \leq A$ and hence,

$$E(m') < E(m).$$

The optimality of the grade assignment m is thus proven.

References

1. L. Bokati, V.V. Kalashnikov, N. Kalashnykova, O. Kosheleva, V. Kreinovich, How to assign grades to tasks so as to maximize student efforts. Russ. Digit. Libr. J. **22**(6), 773–779 (2019)
2. S. Dempe, *Foundations of Bilevel Programming* (Springer Science + Business Media, Dordrecht, 2010)

Part V
Applications to Computing

Most of applications involve intensive computing. In this final Part V, we show that the above-analyzed ideas can be used in all aspects of computing:

- in analyzing the simplest (linear) models (Chap. 24),
- in analyzing more realistic non-linear models (Chap. 25), and even
- in exploring perspective approaches to computing (Chap. 26).

Chapter 24
Why Geometric Progression in Selecting the LASSO Parameter

Most of the applications involve intensive computing. In this and the following chapters, we will show that the ideas of decision making under uncertainty can be used in all aspects of computing. In this chapter, we show that they can be used in analyzing the simplest (linear) models. In the next Chap. 25, we will show that these ideas can help in analyzing non-linear models. Finally, in Chap. 26, we show that these ideas can be useful in the analysis of perspective approaches to computing.

This chapter deals with linear models. For such models, in situations when we know which inputs are relevant, the least squares method is often the best way to solve linear regression problems. However, in many practical situations, we do not know beforehand which inputs are relevant and which are not. In such situations, a 1-parameter modification of the least squares method known as LASSO leads to more adequate results. To use LASSO, we need to determine the value of the LASSO parameter that best fits the given data. In practice, this parameter is determined by trying all the values from some discrete set. It has been empirically shown that this selection works the best if we try values from a geometric progression. In this chapter, we provide a theoretical explanation for this empirical fact.

Comment. Results from this chapter first appeared in [1].

24.1 Formulation of the Problem

Need for regression. In many real-life situations, we know that the quantity y is uniquely determined by the quantities $x_1, \ldots, x_n$, but we do not know the exact formula for this dependence. For example, in physics, we know that the aerodynamic resistance increases with the body's velocity, but we often do not know how exactly.

L. Bokati and V. Kreinovich, *Decision Making Under Uncertainty, with a Special Emphasis on Geosciences and Education*, Studies in Systems, Decision and Control 218,
https://doi.org/10.1007/978-3-031-26086-5_24

In economics, we may know that a change in tax rate influences the economic growth, but we often do not know how exactly.

In all such cases, we need to find the dependence $y = f(x_1, \ldots, x_n)$ between several quantities based on the available data, i.e., based on the previous observations $(x_{k1}, \ldots, x_{kn}, y_k)$ in each of which we know both the values x_{ki} of the input quantities x_i and the value y_k of the output quantity y. In statistics, determining the dependence from the data is known as *regression*.

Need for linear regression. In most cases, the desired dependence is smooth—and usually, it can even be expanded in Taylor series; see, e.g., [2, 3]. In many practical situations, the range of the input variables is small, i.e., we have $x_i \approx x_i^{(0)}$ for some values $x_i^{(0)}$. In such situations, after we expand the desired dependence in Taylor series, we can safely ignore terms which are quadratic or of higher order with respect to the differences $x_i - x_i^{(0)}$ and only keep terms which are linear in terms of these differences:

$$y = f(x_1, \ldots, x_n) = c_0 + \sum_{i=1}^{n} a_i \cdot \left(x_i - x_i^{(0)}\right),$$

where $c_0 \stackrel{\text{def}}{=} f\left(x_1^{(0)}, \ldots, x_n^{(0)}\right)$ and $a_i \stackrel{\text{def}}{=} \dfrac{\partial f}{\partial x_i}_{|x_i = x_i^{(0)}}$. This expression can be simplified into a general linear expression:

$$y = a_0 + \sum_{i=1}^{n} a_i \cdot x_i, \tag{24.1}$$

where $a_0 \stackrel{\text{def}}{=} c_0 - \sum_{i=1}^{n} a_i \cdot x_i^{(0)}$.

In practice, measurements are never absolutely precise, so when we plug in the actually measured values x_{ki} and y_i, we will only get an approximate equality:

$$y_k \approx a_0 + \sum_{i=1}^{m} a_i \cdot x_{ki}. \tag{24.2}$$

Thus, the problem of finding the desired dependence can be reformulated as follows:

- given the values y_k and x_{ki},
- find the coefficients a_i for which the property (24.2) holds for all k.

The usual least squares approach. We want each left-hand side y_k of the approximate equality (24.2) to be close to the corresponding right-hand side. In other words, we want the tuple $(y_1, \ldots, y_K)$ consisting of all the left-hand sides to be close to a similar tuple formed by the right-sides

$$\left(a_0 + \sum_{i=1}^{m} a_i \cdot x_{1i}, \ldots, a_0 + \sum_{i=1}^{m} a_i \cdot x_{Ki}\right).$$

It is reasonable to select the values a_i for which the distance between these two tuples is the smallest possible. Minimizing the distance is equivalent to minimizing the square of this distance, i.e., the expression

$$\sum_{k=1}^{K}\left(y_k - \left(a_0 + \sum_{i=1}^{m} a_i \cdot x_{ki}\right)\right)^2. \tag{24.3}$$

This minimization is known as the *Least Squares method*. This is the most widely used method for processing data. The corresponding values a_i can be easily found if we differentiate the quadratic expression (24.3) with respect to each of the unknowns a_i and then equate the corresponding linear expressions to 0. Then, we get an easy-to-solve system of linear equations.

Comment. The above heuristic idea becomes well-justified when we consider the case when the measurement errors are normally distributed with 0 mean and the same standard deviation σ. This indeed happens in many situations when the measuring instrument's bias has been carefully eliminated, and most major sources of measurement errors have been removed. In such situations, the resulting measurement error is a joint effect of many similarly small error components. For such joint effects, the Central Limit Theorem states that the resulting distribution is close to Gaussian (=normal); see, e.g., [4]. Once we know the probability distributions, a natural idea is to select the most probable values a_i, i.e., the values for which the probability to observe the values y_k is the largest. For normal distributions, this idea leads exactly to the least squares method.

Need to go beyond least squares. When we know that all the inputs x_i are essential to predict the value y of the desired quantity, the least squares method works reasonably well. The problem is that in practice, we often do not know which inputs x_i are relevant and which are not. As a result, to be on the safe side, we include as many inputs as possible, perfectly understanding that many of them will turn out to be irrelevant.

If all the measurements were exact, this would not be a problem: for irrelevant inputs x_i, we would get $a_i = 0$, and the resulting formula would be the desired one. However, because of the measurement errors, we do not get exactly 0s. Moreover, the more such irrelevant variables we add, the more non-zero "noise" terms $a_i \cdot x_i$ we will have, and the larger will be their sum—negatively affecting the accuracy of the formula (24.3) and thus, of the resulting desired (non-zero) coefficients a_i.

LASSO method. Since we know that many coefficients will be 0, a natural idea is, instead of considering all possible tuples $a \stackrel{\text{def}}{=} (a_0, a_1, \ldots, a_n)$, to only consider tuples for which a bounded number of coefficients is 0, i.e., for which $\|a\|_0 \le B$ for some constant b, where $\|a\|_0$ (known as the ℓ_0-norm) denotes the number of non-zero coefficients in a tuple.

The problem with this natural idea is that the resulting optimization problem becomes NP-hard, which means, crudely speaking, that no feasible algorithm is possible that would always solve all the instances of this problem. A usual way to solve such problem is by replacing the ℓ_0-norm with an ℓ_1-norm $\sum_{i=0}^{n} |a_i|$ which is convex and for which, therefore, the optimization problem is easier to solve. So, instead of solving the problem of unconditionally minimizing the expression (24.3), we minimize this expression under the constraint $\sum_{i=0}^{n} |a_i| \leq B$ for some constant B. This minimum can be attained when we have strict inequality or when the constraint becomes an equality. If the constraint is a strict inequality, then we have a local minimum of (24.3), which, for quadratic functions, is exactly the global minimum that we try to avoid. Thus, to avoid using least squares, we must consider the case when the constraint becomes an equality $\sum_{i=0}^{n} |a_i| = B$.

According to the Lagrange multiplier method, minimizing a function under an equality-type constraint is equivalent, for an appropriate value of a parameter λ, to unconstrained minimization of the linear combination of the original objective function and the constraint, i.e., to minimizing the expression

$$\sum_{k=1}^{K} \left(y_k - \left(a_0 + \sum_{i=1}^{m} a_i \cdot x_{ki} \right) \right)^2 + \lambda \cdot \sum_{i=0}^{n} |a_i|. \tag{24.4}$$

This minimization is known as the *Least Absolute Shrinkage and Selection Operator* method—*LASSO method*, for short; see, e.g., [5, 6].

How the LASSO parameter λ is selected: main idea. The success of the LASSO method depends on what value λ we select. When λ is close to 0, we retain all the problems of the usual least squares method. When λ is too large, the λ-term dominates, so we select the values $a_i = 0$, which do not provide any good description of the desired dependence.

In different situations, different values λ will work best. The more irrelevant inputs we have, the more important it is to deviate form the least squares, and thus, the larger the parameter λ—that describes this deviation—should be. We rarely know beforehand which inputs are relevant—this is the whole problem—so we do now know beforehand what value λ we should use. The best value λ needs to be decided based on the data.

A usual way of testing any dependence is by randomly dividing the data into a (larger) training set and a (smaller) testing set. We use the training set to find the value of the desired parameters (in our case, the parameters a_i), and then we use the testing set to gauge how good is the model. As usual with the methods using randomization, to get more reliable results, we can repeat this procedure several times, and make sure that the results are good for all cases,

In precise terms, we select several training subsets $S_1, \ldots, S_m \subseteq \{1, \ldots, K\}$. For each of these subsets S_j, we find the values $a_{ji}(\lambda)$ that minimize the functional

$$\sum_{k \in S_j} \left(y_k - \left(a_0 + \sum_{i=1}^{m} a_i \cdot x_{ki} \right) \right)^2 + \lambda \cdot \sum_{i=0}^{n} |a_i|. \tag{24.5}$$

We can then compute the overall inaccuracy, as

$$\Delta(\lambda) \stackrel{\text{def}}{=} \sum_{j=1}^{m} \left(\sum_{k \notin S_j} \left(y_k - \left(a_{j0}(\lambda) + \sum_{i=1}^{m} a_{ji}(\lambda) \cdot x_{ki} \right) \right)^2 \right). \tag{24.6}$$

We then select the value λ for which the corresponding inaccuracy is the smallest possible.

How the LASSO parameter λ is selected: details. In the ideal world, we should be able to try all possible real values λ. However, there are infinitely many real numbers, and in practice, we can only test finitely many of them. Which set of values λ should we choose?

It turned out that empirically, the best results are obtained of we use the values λ that form a geometric progression $\lambda_n = c_0 \cdot q^n$. Of course, a geometric progression also has infinitely many values, but we do not need to test all of them: usually, as λ increases from 0, the value $\Delta(\lambda)$ first decreases then increases again, so it is enough to catch a moment when this value starts increasing.

Natural question and what we do in this chapter. A natural question is: why geometric progression works best? In this chapter, we provide a theoretical explanation for this empirical fact.

24.2 Our Result

What do we want? At first glance, the answer to this question is straightforward: we want to select a discrete set of values, i.e., a set

$$S = \{\ldots < \lambda_n < \lambda_{n+1} < \ldots\}.$$

However, a deeper analysis shows that the answer is not so simple. Indeed, what we are interested in is the dependence between the quantities y and x_i. However, what we have to deal with is not the quantities themselves, but their numerical values, and the numerical values depend on what unit we choose for measuring these quantities. For example:

- a person who is 1.7 m high is also 170 cm high,
- an April 2020 price of 2 US dollars is the same as the price of $2 \cdot 23500 = 47000$ Vietnam Dong, etc.

In most cases, the choice of the units is rather arbitrary. It is therefore reasonable to require that the results of data processing—when converted to original units—should not depend on which units we originally used. And hereby lies a problem. Suppose that we keep the same units for x_i but change a measuring unit for y to a one which is α times smaller. In this case, the new numerical values of y become α times larger: $y \to y' = \alpha \cdot y$. To properly capture these new values, we need to increase the original values a_i by the same factor, i.e., replace the values a_i with the new values $a'_i = \alpha \cdot a_i$. In terms of these new values, the minimized expression (24.4) takes the form

$$\sum_{k=1}^{K} \left(y'_k - \left(a'_0 + \sum_{i=1}^{m} a'_i \cdot x_{ki} \right) \right)^2 + \lambda \cdot \sum_{i=0}^{n} |a'_i|,$$

i.e., taking into account that $y'_k = \alpha \cdot y_k$ and $a'_i = \alpha \cdot a_i$, the form

$$\alpha^2 \cdot \sum_{k=1}^{K} \left(y_k - \left(a_0 + \sum_{i=1}^{m} a_i \cdot x_{ki} \right) \right)^2 + \alpha \cdot \lambda \cdot \sum_{i=0}^{n} |a_i|.$$

Minimizing an expression is the same as minimizing α^{-2} times this expression, i.e., the modified expression

$$\sum_{k=1}^{K} \left(y_k - \left(a_0 + \sum_{i=1}^{m} a_i \cdot x_{ki} \right) \right)^2 + \alpha^{-1} \cdot \lambda \cdot \sum_{i=0}^{n} |a_i|.$$

This new expression is similar to the original minimized expression (24.4), but with a new value of the LASSO parameter $\lambda' = \alpha^{-1} \cdot \lambda$.

What this says is that when we change the measuring units, the values of λ are also re-scaled—i.e., multiplied by a constant. What was the set $\{\lambda_n\}$ in the old units becomes the re-scaled set $\{\alpha^{-1} \cdot \lambda_n\}$ in the new units. Since this is, in effect, the same set but corresponding to different measuring units, we cannot say that one of these sets is better than the other, they clearly have the same quality.

So, we cannot choose a single set S, we must choose a family of sets $\{c \cdot S\}_c$, where the notation $c \cdot S$ means $\{c \cdot \lambda : \lambda \in S\}$.

Natural uniqueness requirement. Eventually, we need to select some set S. As we have just explained, we cannot select one set a priori, since with every set S, a set $c \cdot S$ also has the same quality. To fix a unique set, we can, e.g., fix one of the values $\lambda \in S$. Let us require that with this fixture, we will be end up with a unique optimal set S. This means, in particular, that, if we select a real number $\lambda \in S$, then the only set $c \cdot S$ that contains this number will be the same set S.

Let us describe this requirement in precise terms.

Definition 24.1 By a *discrete set*, we mean a subset S of the set $\mathbb{R}^+$ of all positive real numbers for which, for every $\lambda \in S$, there exists an $\varepsilon > 0$ such that for every other element $\lambda' \in S$, we have $|\lambda - \lambda'| > \varepsilon$.

Comment. For such sets, for each element λ, if there are larger elements, then there is the "next" element—i.e., the smallest element which is larger than λ. Similarly, if there are smaller elements, then there exists the "previous" element—i.e., the largest element which is smaller than λ. Thus, such sets have the form $\{\ldots < \lambda_{n-1} < \lambda_n < \lambda_{n+1} < \ldots\}$

Notation. For each set S and for each number $c > 0$, by $c \cdot S$, we mean the set

$$\{c \cdot \lambda : \lambda \in S\}.$$

Definition 24.2 We say that a discrete set S is *uniquely determined* if for every $\lambda \in S$ and $c > 0$, if $\lambda \in c \cdot S$, then $c \cdot S = S$.

Proposition 24.1 *A set S is uniquely determined if and only if it is a geometric progression, i.e., if and only if it has the form*

$$S = \{c_0 \cdot q^n : n = \ldots, -2, -1, 0, 1, 2, \ldots\}$$

for some c_0 and q.

Discussion. This results explains why geometric progression is used to select the LASSO parameter λ.

Proof It is easy to prove that every geometric progression is uniquely determined. Indeed, if for $\lambda = c_0 \cdot q^n$, we have $\lambda \in c \cdot S$, this means that $\lambda = c \cdot c_0 \cdot q^m$ for some m, i.e., $c_0 \cdot q^n = c \cdot c_0 \cdot q^m$. Dividing both sides by $c_0 \cdot q^m$, we conclude that $c = q^{n-m}$ for some integer $n - m$. Let us show that in this case, $c \cdot S = S$. Indeed, each element x of the set $c \cdot S$ has the form $x = c \cdot c_0 \cdot q^k$ for some integer k. Substituting $c = q^{n-m}$ into this formula, we conclude that $x = c_0 \cdot q^{k+(n-m)}$, i.e., that $x \in S$. Similarly, we can prove that if $x \in S$, then $x \in c \cdot S$.

Vice versa, let us assume that the set S is uniquely determined. Let us pick any element $\lambda \in S$ and denote it by λ_0. The next element we will denote by λ_1, the next to next by λ_2, etc. Similarly, the element previous to λ_0 will be denoted by λ_{-1}, previous to previous by λ_{-2}, etc. Thus,

$$S = \{\ldots, \lambda_{-2}, \lambda_{-1}, \lambda_0, \lambda_1, \lambda_2, \ldots\}.$$

Clearly, $\lambda_1 \in S$, and for $q \stackrel{\text{def}}{=} \lambda_1/\lambda_0$, we have $\lambda_1 \in q \cdot S$—since $\lambda_1 = (\lambda_1/\lambda_0) \cdot \lambda_0 = q \cdot \lambda_0$ for $\lambda_0 \in S$. Since the set S is uniquely determined, this implies that $q \cdot S = S$. Since

$$S = \{\ldots, \lambda_{-2}, \lambda_{-1}, \lambda_0, \lambda_1, \lambda_2, \ldots\},$$

we have

$$q \cdot S = \{\ldots, q \cdot \lambda_{-2}, q \cdot \lambda_{-1}, q \cdot \lambda_0, q \cdot \lambda_1, q \cdot \lambda_2, \ldots\}.$$

The sets S and $q \cdot S$ coincide. We know that $q \cdot \lambda_0 = \lambda_1$. Thus, the element next to $q \cdot \lambda_0$ in the set $q \cdot S$—i.e., the element $q \cdot \lambda_1$—must be equal to the element which is next to λ_1 in the set S, i.e., to the element λ_2: $\lambda_2 = q \cdot \lambda_1$. For next to next elements, we get $\lambda_3 = q \cdot \lambda_2$ and, in general, we get $\lambda_{n+1} = q \cdot \lambda_n$ for all n—which is exactly the definition of a geometric progression.

The proposition is proven.

Comment. Similar arguments can explain why in machine learning methods such as deep learning (see, e.g., [7])—which usually use the gradient method $x_{i+1} = x_i - \lambda_i \cdot \dfrac{\partial J}{\partial x_i}$ to find the minimum of an appropriate objective function J, empirically the best strategy for selecting λ_i also follows approximately a geometric progression: e.g., some algorithms use:

- $\lambda_i = 0.1$ for the first ten iterations,
- $\lambda_i = 0.01$ for the next ten iterations,
- $\lambda_i = 0.001$ for the next ten iterations, etc.

Indeed, in this case, similarly, re-scaling of J is equivalent to re-scaling of λ, and thus, we need to have a family of sequences $\{c \cdot \lambda_i\}$ corresponding to different $c > 0$. A natural uniqueness requirement—as we have shown—leads to the geometric progression.

References

1. W. Kubin, Y. Xie, L. Bokati, V. Kreinovich, K. Autchariyapanitkul, Why geometric progression in selecting the LASSO parameter: a theoretical explanation, University of Texas at El Paso, Department of Computer Science, Technical Report UTEP-CS-20-35, Apr. 2020
2. R. Feynman, R. Leighton, M. Sands, *The Feynman Lectures on Physics* (Addison Wesley, Boston, MA, 2005)
3. K.S. Thorne, R.D. Blandford, *Modern Classical Physics: Optics, Fluids, Plasmas, Elasticity, Relativity, and Statistical Physics* (Princeton University Press, Princeton, NJ, 2017)
4. D.J. Sheskin, *Handbook of Parametric and Non-parametric Statistical Procedures* (Chapman & Hall/CRC, London, UK, 2011)
5. T. Hastie, R. Tibshirani, M. Wainwright, *Statistical Learning with Sparsity: The Lasso and Generalizations* (Chapman & Hall/CRC, Boca Raton, FL, 2015)
6. R. Tibshirani, Regression shrinkage and selection via the LASSO. J. R. Stat. Soc. **58**(1), 267–288 (1996)
7. I. Goodfellow, Y. Bengio, A. Courville, *Deep Learning* (MIT Press, Cambridge, MA, 2016)

Chapter 25
Applications to Computing: Why Deep Learning is More Efficient Than Support Vector Machines, and How It is Related to Sparsity Techniques in Signal Processing

In the previous chapter, we showed how ideas of decision making under uncertainty can be applied to the case of linear dependence. In this chapter, we analyze a general case of possible non-linear dependence. Using computers to find such a generic dependence is what constitutes machine learning. Several decades ago, traditional neural networks were the most efficient machine learning technique. Then it turned out that, in general, a different technique called support vector machines is more efficient. Reasonably recently, a new technique called deep learning has been shown to be the most efficient one. These are empirical observations, but how we explain them—thus making the corresponding conclusions more reliable? In this chapter, we provide a possible theoretical explanation for the above-described empirical comparisons. This explanation enables us to explain yet another empirical fact—that sparsity techniques turned out to be very efficient in signal processing.

Comment. Results from this chapter first appeared in [1].

25.1 Problem Formulation

Main objectives of science and engineering. We want to make our lives better, we want to select actions and designs that will make us happier, we want to improve the world so as to increase our happiness level. To do that we need to know what is the current state of the world, and what changes will occur if we perform different actions. Crudely speaking, learning the state of the world and learning what changes will happen is the main objective of science, while using this knowledge to come up with the best actions and best designs is the main objective of engineering.

L. Bokati and V. Kreinovich, *Decision Making Under Uncertainty, with a Special Emphasis on Geosciences and Education*, Studies in Systems, Decision and Control 218,
https://doi.org/10.1007/978-3-031-26086-5_25

Need for machine learning. In some cases, we already know how the world operates: e.g., we know that the movement of the celestial bodies is well described by Newton's equations—it is described so well that we can predict, e.g., Solar eclipses centuries ahead. In many other cases, however, we do not have such a good knowledge, we need to extract the corresponding laws of nature from the observations.

In general, prediction means that we can predict the future value y of the physical quantity of interest based on the current and past values $x_1, \ldots, x_n$ of related quantities. To be able to do that, we need to have an algorithm that, given the values $x_1, \ldots, x_n$, computes a reasonable estimate for the desired future value y.

In the past, designing such algorithms was done by geniuses—Newton described how to predict the motion of celestial bodies, Einstein provided more accurate algorithms, Schroedinger, in effect, described how to predict probabilities of different states of the quantum system, etc. This still largely remains the domain of geniuses, Nobel prizes are awarded every year for these discoveries. However, now that the computers have become very efficient, they are often used to help. This use of computers is known as *machine learning*: when we know, in several cases $c = 1, \ldots, C$, which values $y^{(c)}$ corresponded to appropriate values $x_1^{(c)}, \ldots, x_n^{(c)}$, and we want to find an algorithm $f(x_1, \ldots, x_n)$ for which, for all these cases c, we have $y^{(c)} \approx f(x_1^{(c)}, \ldots, x_n^{(c)})$.

The value y may be tomorrow's temperature in a given area, it may be a binary (0–1) variable deciding whether a given email is legitimate or a spam (or whether, e.g., the given image is an image of a cat).

Machine learning: a brief history. One of the first successful general machine learning techniques was the technique of *neural networks*; see, e.g., [2]. In this technique, we look for algorithms of the type

$$f(x_1, \ldots, x_n) = \sum_{k=1}^{K} W_k \cdot s\left(\sum_{i=1}^{n} w_{ki} \cdot x_i - w_{k0}\right) - W_0,$$

for some non-linear function $s(z)$ called an *activation function*, and for some values w_{ki} and W_k knows as *weights*. As the function $s(z)$, researchers usually selected the so-called *sigmoid function*

$$s(z) = \frac{1}{1 + \exp(-z)}.$$

This algorithm emulates a 3-layer network of biological neurons—the main cells providing data processing in our brains. In the first layer, we have input neurons that read the inputs $x_1, \ldots, x_n$. In the second layer—called a *hidden layer*—we have K neurons each of which first generates a linear combination

$$z_k = \sum_{i=1}^{n} w_{ki} \cdot x_i - w_{k0}$$

of the input signals, and then applies an appropriate nonlinear function $s(z)$ to this combination, resulting in a signal $y_k = s(z_k)$. The processing by biological neurons is well described by the sigmoid activation function—this is the reason why this function was selected for artificial neural networks in the first place. After that, in the final output layer, the signals y_k from the neurons in the hidden layer are combined into a linear combination $\sum\limits_{k=1}^{K} W_k \cdot y_k - W_0$ which is returned as the output.

A special efficient algorithm—known as *backpropagation*—was developed to *train* the corresponding neural network, i.e., to find the values of the weights that provide the best fit for the observation results $x_1^{(c)}, \ldots, x_n^{(c)}, y^{(c)}$.

Support Vector Machines: a brief description. Later, in many practical problems, a different technique became more efficient: the technique of *Support Vector Machines*; see, e.g., [3] and references therein. Let us explain this technique on the example of a *binary classification* problem, i.e., a problem in which we need to classify all objects (or events) into one of two classes, based on the values $x_1, \ldots, x_n$ of the corresponding parameters—i.e., in which the desired output y has only two possible values.

In general, if, based on the values $x_1, \ldots, x_n$ we can uniquely determine to which of the two classes this object belongs, this means that the set of all possible values of the tuple $x = (x_1, \ldots, x_n)$ is divided into two non-intersecting sets S_1 and S_2 corresponding to each of the two classes.

We can therefore come up with a continuous function $f(x_1, \ldots, x_n)$ such that $f(x) \geq 0$ for $x \in S_1$ and $f(x) \leq 0$ for $x \in S_2$. As an example of such a function, we can take $f(x) = d(x, S_2) - d(x, S_1)$, where the distance $d(x, S)$ between a point x and the set S is defined as the distance from x to the closest point of S, i.e., as $\inf\limits_{s \in S} d(x, s)$. Clearly, if $x \in S$, then $d(x, s) = 0$ for $s = x$ thus $d(x, S) = 0$.

- For points $x \in S_1$, we have $d(x, S_1) = 0$ but usually $d(x, S_2) > 0$, thus $f(x) = d(x, S_2) - d(x, S_1) > 0$.
- On the other hand, for points $x \in S_2$, we have

$$d(x, S_2) = 0$$

while, in general, $d(x, S_1) > 0$, thus

$$f(x) = d(x, S_2) - d(x, S_1) < 0.$$

In some simple cases, there exists a linear function

$$f(x_1, \ldots, x_n) = a_0 + \sum_{i=1}^{n} a_i \cdot x_i$$

that separates the two classes. In this case, there exist efficient algorithms for finding the corresponding coefficients a_i—for example, we can use linear programming (see, e.g., [4, 5]) to find the values a_i for which:

- $a_0 + \sum\limits_{i=1}^{n} a_i \cdot x_i > 0$ for all known tuples $x \in S_1$, and
- $a_0 + \sum\limits_{i=1}^{n} a_i \cdot x_i < 0$ for all known tuples $x \in S_2$.

In many practical situations, however, such a linear separation is not possible. In such situations, we can take into account the known fact that any continuous function on a bounded domain (and for practical problems, there are always bounds on the values of all the quantities) can be approximated, with any given accuracy, by a polynomial. Thus, with any given accuracy, we can separate the two classes by checking whether the f-approximating polynomial

$$P_f(x) = a_0 + \sum_{i=1}^{n} a_i \cdot x_i + \sum_{i=1}^{n} \sum_{j=1}^{n} a_{ij} \cdot x_i \cdot x_j + \cdots$$

is positive or negative.

In other words, if we perform a non-linear mapping of each original n-dimensional point $x = (x_1, \ldots, x_n)$ into a higher-dimensional point

$$X = (X_1, \ldots, X_n, X_{11}, X_{12}, \ldots, X_{nn}, \ldots) =$$

$$(x_1, \ldots, x_n, x_1^2, x_1 \cdot x_2, \ldots, x_n^2, \ldots),$$

then in this higher-dimensional space, the separating function becomes linear:

$$P_f(X) = a_0 + \sum_{i=1}^{n} a_i \cdot X_i + \sum_{i=1}^{n} \sum_{j=1}^{n} a_{ij} \cdot X_{ij} + \cdots,$$

and we know how to effectively find a linear separation.

Instead of polynomials, we can use another basis $e_1(x), e_2(x), \ldots$, to approximate a general separating function as

$$a_1 \cdot e_1(x) + a_2 \cdot e_2(x) + \cdots$$

The name of this technique comes from the fact that when solving the corresponding linear programming problem, we can safely ignore many of the samples and concentrate only on the vectors X which are close to the boundary between the two sets—if we get linear separation for such *support vectors*, we will automatically get separation for other vectors X as well.

This possibility to decrease the number of iterations enables us to come up with algorithms for the SVM approach which are more efficient than general linear programming algorithms—and many other ideas and tricks help make the resulting algorithms even faster.

Deep learning: a brief description. Lately, the most efficient machine learning tool is *deep learning*; see, e.g., [6]. Deep learning is a version of a neural network, but the main difference is that instead of a large number of neurons in a hidden layer, we have multiple layers with a relatively small number of neurons in each of them.

Similarly to the traditional neural networks, we start with the inputs $x_1, \ldots, x_n$. These inputs serve as inputs $x_i^{(0)}$ to the neurons in the first later. On each layer k, each neuron takes, as inputs, outputs $x_i^{(k-1)}$ from the previous layer and returns the value

$$x_j^{(k)} = s_k\left(\sum_i w_{ij}^{(k)} \cdot x_i^{(k-1)} - w_{0j}^{(k)}\right).$$

For most layers, instead of the sigmoid, it turns out to be more efficient to use a piece-wise linear function $s_k(x) = \max(x, 0)$ which is:

- equal to 0 for $x < 0$ and
- equal to x for $x > 0$.

In the last layer, sometimes, the sigmoid is used.

There are also layers in which inputs are divided into groups, and we combine inputs from each group into a single value—e.g., by taking the maximum of the corresponding values.

In addition to the general backpropagation idea, several other techniques are used to speed up the corresponding computations—e.g., instead of using *all* the neurons in training, one of the techniques is to only use, on each iteration, *some* of the neurons and then combine the results by applying an appropriate combination functions (which turns out to be geometric mean).

Natural questions. So far, we have described what happened: support vector machines turned out to be more efficient in machine learning, and deep learning is, in general, more efficient than support vector machines. A natural question is: why? How can we theoretically explain these empirical facts—thus increasing our trust in the corresponding conclusions?

What we do in this chapter. In general deep learning is more efficient than the traditional neural networks; see, e.g., [7–9]. In this chapter, we extend these explanations to the comparison between support vector machines and neural networks.

The resulting explanation will help us understand yet another empirical fact—the empirical efficiency of sparse techniques in signal processing.

25.2 Support Vector Machines Versus Neural Networks

This empirical comparison is the easiest to explain. Indeed, to train a traditional neural network on the given cases $x_1^{(c)}, \ldots, x_n^{(c)}, y^{(c)}$, we need to find the weights W_k and w_{ki} for which

$$y^{(c)} \approx \sum_{k=1}^{K} W_k \cdot s\left(\sum_{i=1}^{n} w_{ki} \cdot x_i^{(c)} - w_{k0}\right) - W_0.$$

Here, the activation function $s(z)$ is non-linear, so we have a system of non-linear equations for finding the corresponding weights W_k and w_{ki}. In general, solving a system of nonlinear equations is NP-hard even for quadratic equations; see, e.g., [10, 11].

In contrast, for support vector machines, to find the corresponding coefficients a_i, it is sufficient to solve a linear programming problem—and this can be done in feasible time. This explains why support vector machines are more efficient than traditional neural networks.

25.3 Support Vector Machines Versus Deep Learning

At first glance, the above explanation should work for the comparison between support vector machines and deep networks: in the first case, we have a feasible algorithm, while in the second case, we have an NP-hard problem that may require very long (exponential) time.

However, this is only at first glance. Namely:

- the above comparison assumes that all the inputs $x_1, \ldots, x_n$ are independent—in the sense of functional dependency, i.e., that none of them can be described in terms of one another.
- In reality, most inputs are dependent in this sense.

This is especially clear in many engineering and scientific applications, where we use the results of measuring appropriate quantities at different moments of time as inputs for prediction, and we know that these quantities are usually *not* independent—they satisfy some differential equations. As a result, we do not need to use all n inputs. If there are $m \ll n$ independent ones, this means that it is sufficient to use only m of the inputs—or, alternatively, m different combinations of inputs, as long as these combinations are independent (and, in general, they are); see, e.g., [12].

And this is exactly what is happening in a deep neural network. Indeed, in the traditional neural network, in which we have many neurons in the processing (hidden) layer—we can have as many neurons as inputs (or even more). In contrast, in the deep neural networks, the number of neurons in each layer is limited. In particular, the number of neurons in the first processing layer is, in general, much smaller than the number of inputs. And all the resulting computations are based *only* on the outputs $x_k^{(1)}$ of the neurons from this first layer. Thus, in effect, the desired quantity y is computed not based on all n inputs, but based only on m combinations—where m is the number of neurons in the first processing layer.

The empirical fact—that, in spite of this limitation, deep neural networks seem to provide a universal approximation to all kinds of actual dependencies—is an indication that, inputs are usually dependent on each other.

This dependence explains why, empirically, deep neural networks work better than support vector machines—deep networks implicitly take into account this dependency, while support vector machines do not take any advantage of this dependency. As a result, deep networks need fewer parameters than would be needed if they would consider n functionally independent inputs. Hence, during the same time, they can perform more processing and thus, get more accurate predictions.

Comment. In this chapter, we provide a possible theoretical explanation for the fact that support vector machines are, on average, more efficient than traditional neural networks but less efficient than deep learning. To make our theoretical explanations more convincing, it is desirable to have additional experimental data supporting these explanations.

25.4 Sparsity Techniques: An Explanation of Their Efficiency

What are sparsity techniques. The above explanations help us explain another empirical fact: that in many applications of signal and image processing, sparsity techniques have been very effective. Specifically, usually, in signal processing, we represent the signal $x(t)$ by the coefficients a_i of its expansion in the appropriate basis $e_1(t), e_2(t)$, etc.: $x(t) \approx \sum\limits_{i=1}^{n} a_i \cdot e_i(t)$;

- in Fourier analysis, we use the basic of sines and cosines;
- in wavelet analysis, we use wavelets as the basis, etc.

Similarly, in image processing, we represent an image $I(x)$ by the coefficients of its expansion over some basis.

It turns out that in many practical problems, we can select the basis $e_i(t)$ in such a way that for most actual signals, the corresponding representation becomes *sparse* in the sense that most of the corresponding coefficients a_i are zeros. This phenomenon leads to very efficient algorithms for signal and image processing; see, e.g., [13–29]. However, while empirically successful, from the theoretical viewpoint, this phenomenon largely remains a mystery: why can we find such a basis? Some preliminary explanations are provided in the papers [30, 31], but additional explanations are definitely desirable.

Our new explanation. The shape of the actual signal $x(t)$ depends on many different phenomena. So, in general, we can say that

$$x(t) = F(t, c_1, \ldots, c_N)$$

for some function F, where $c_1, \ldots, c_N$ are numerical values characterizing all these phenomena.

Usual signal processing algorithms implicitly assume that we can have all possible combinations of these values c_i. However, as we have mentioned, in reality, the corresponding phenomena are dependent on each other. As a result, there is a functional dependence between the corresponding values c_i. Only few of them $m \ll N$ are truly independent, others can be determined based on these few ones.

If we denote the corresponding m independent values by $b_1, \ldots, b_m$, then the above description takes the form

$$x_i(t) = G(t, b_1, \ldots, b_m)$$

for an appropriate function G.

It is known that any continuous function—in particular, our function G—can be approximated by piecewise linear functions. If we use this approximation instead of the original function G, then we conclude that the domain of possible values of the tuples $(b_1, \ldots, b_m)$ is divided into a small number of sub-domains $D_1, \ldots, D_p$ on each of which D_j the dependence of $x_i(t)$ on the values b_i is linear, i.e., has the form

$$x_i(t) = \sum_{k=1}^{m} b_k \cdot e_{jk}(t),$$

for some functions $e_{jk}(t)$.

So, if we take all $m \cdot p$ the functions $e_{jk}(t)$ corresponding to different subdomains as the basis, we conclude that on each subdomain, each signal can be described by no more than $m \ll p \cdot m$ non-zero coefficients—this is exactly the phenomenon that we observe and utilize in sparsity techniques.

References

1. L. Bokati, O. Kosheleva, V. Kreinovich, A. Sosa, Why deep learning is more efficient than support vector machines, and how it is related to sparsity techniques in signal processing, in *Proceedings of the 2020 4th International Conference on Intelligent Systems, Metaheuristics & Swarm Intelligence ISMSI'2020*, Thimpu, Bhutan, 18–19 Apr. 2020
2. C.M. Bishop, *Pattern Recognition and Machine Learning* (Springer, New York, 2006)
3. I. Steinwart, A. Christmann, *Support Vector Machines* (Springer, New York, 2008)

4. Th.H. Cormen, C.E. Leiserson, R.L. Rivest, C. Stein, *Introduction to Algorithms* (MIT Press, Cambridge, MA, 2009)
5. R.J. Vanderbei, *Linear Programming: Foundations and Extensions* (Springer, New York, 2014)
6. I. Goodfellow, Y. Bengio, A. Courville, *Deep learning* (MIT Press, Cambridge, MA, 2016)
7. C. Baral, O. Fuentes, V. Kreinovich, Why deep neural networks: a possible theoretical explanation, in *Constraint Programming and Decision Making: Theory and Applications*, ed. by M. Ceberio, V. Kreinovich (Springer, Berlin, Heidelberg, 2018), pp. 1–6
8. Kreinovich, V.: From traditional neural networks to deep learning: towards mathematical foundations of empirical successes, in Proceedings of the World Conference on Soft Computing, Baku, Azerbaijan, ed. by S.N. Shahbazova et al. (2018)
9. V. Kreinovich, O. Kosheleva, *Optimization under Uncertainty Explains Empirical Success of Deep Learning Heuristics*, University of Texas at El Paso, Department of Computer Science, Technical Report UTEP-CS-19-49, 2019. http://www.cs.utep.edu/vladik/2019/tr19-49.pdf
10. V. Kreinovich, A. Lakeyev, J. Rohn, P. Kahl, *Computational Complexity and Feasibility of Data Processing and Interval Computations* (Kluwer, Dordrecht, 1998)
11. C. Papadimitriou, *Computational Complexity* (Addison-Wesley, Reading, MA, 1994)
12. V. Kreinovich, O. Kosheleva, J. Urenda, Why such a nonlinear process as protein synthesis is well approximated by linear formulas, in *Abstracts of the 2019 Southwest and Rocky Mountain Regional Meeting of the American Chemical Society SWRMRM'2019*, El Paso, TX, 13–16 Nov. 2019
13. B. Amizic, L. Spinoulas, R. Molina, A.K. Katsaggelos, Compressive blind image deconvolution. IEEE Trans. Image Process. **22**(10), 3994–4006 (2013)
14. E.J. Candès, J. Romberg, T. Tao, Stable signal recovery from incomplete and inaccurate measurements. Commun. Pure Appl. Math. **59**, 1207–1223 (2006)
15. E. Candès, J. Romberg, T. Tao, Robust uncertainty principles: exact signal reconstruction from highly incomplete frequency information. IEEE Trans. Inf. Theory **52**(2), 489–509 (2006)
16. E.J. Candès, T. Tao, Decoding by linear programming. IEEE Trans. Inf. Theory **51**(12), 4203–4215 (2005)
17. E.J. Candès, M.B. Wakin, An Introduction to compressive sampling. IEEE Signal Process. Mag. **25**(2), 21–30 (2008)
18. D.L. Donoho, Compressed sensing. IEEE Trans. Inf. Theory **52**(4), 1289–1306 (2005)
19. M.F. Duarte, M.A. Davenport, D. Takhar, J.N. Laska, T. Sun, K.F. Kelly, R.G. Baraniuk, Single-pixel imaging via compressive sampling. IEEE Signal Process. Mag. **25**(2), 83–91 (2008)
20. T. Edeler, K. Ohliger, S. Hussmann, A. Mertins, Super-resolution model for a compressed-sensing measurement setup. IEEE Trans. Instrum. Meas. **61**(5), 1140–1148 (2012)
21. M. Elad, *Sparse and Redundant Representations* (Springer, 2010)
22. Y.C. Eldar, G. Kutyniok (eds.), *Compressed Sensing: Theory and Applications* (Cambridge University Press, New York, 2012)
23. S. Foucart, H. Rauhut, *A Mathematical Introduction to Compressive Sensing* (Birkhäuser, New York, 2013)
24. J. Ma, F.-X. Le Dimet, Deblurring from highly incomplete measurements for remote sensing. IEEE Trans. Geosci. Remote Sens. **47**(3), 792–802 (2009)
25. L. McMackin, M.A. Herman, B. Chatterjee, M. Weldon, A high-resolution SWIR camera via compressed sensing. Proc. SPIE **8353**(1), 03–8353 (2012)
26. B.K. Natarajan, Sparse approximate solutions to linear systems. SIAM J. Comput. **24**, 227–234 (1995)
27. V.M. Patel, R. Chellappa, *Sparse Representations and Compressive Sensing for Imaging and Vision* (Springer, New York, 2013)
28. Y. Tsaig, D. Donoho, Compressed sensing. IEEE Trans. Inform. Theory **52**(4), 1289–1306 (2006)
29. L. Xiao, J. Shao, L. Huang, Z. Wei, Compounded regularization and fast algorithm for compressive sensing deconvolution, in *Proceedings of the 6th International Conference on Image Graphics* pp. 616–621 (2011)

30. F. Cervantes, B. Usevitch, L. Valera, V. Kreinovich, Why sparse? Fuzzy techniques explain empirical efficiency of sparsity-based data- and image-processing algorithms, in *Recent Developments and New Direction in Soft Computing: Foundations and Applications*. ed. by L. Zadeh, R.R. Yager, S.N. Shahbazova, M. Reformat, V. Kreinovich (Springer, Cham, Switzerland, 2018), pp. 19–430
31. T. Dumrongpokaphan, O. Kosheleva, V. Kreinovich, A. Belina, Why sparse?, in *Beyond Traditional Probabilistic Data Processing Techniques: Interval, Fuzzy, etc*,ed. by O. Kosheleva, S. Shary, G. Xiang, R. Zapatrin (Methods and Their Applications, Springer, Cham, Switzerland, 2020), pp. 61–470

Chapter 26
Applications to Computing: Representing Functions in Quantum and Reversible Computing

In the previous two chapters, we showed how ideas of decision making under uncertainty can be applied to the current data processing algorithms. However, for many practical problems, the existing algorithms take too long a time. It is therefore necessary to explore the possibility of faster computations. One such perspective possibility is the use of quantum computing. There are many existing algorithms for quantum computing, but, as we show in this chapter, they often do not provide an adequate representation of generic functions—and objective functions corresponding to decision making under uncertainty can be very complex. In this chapter, we show how to come up with more adequate representation of generic functions in quantum computing.

Comment. Results from this chapter first appeared in [1].

26.1 Formulation of the Problem

Need for faster computing. While computers are very fast, in many practical problems, we need even faster computations. For example, we can, in principle, with high accuracy predict in which direction a deadly tornado will turn in the next 15 min, but this computation requires hours even on the most efficient high performance computers—too late for the resulting prediction to be of any use.

Faster computations means smaller processing units. One of the main limitations on physical processes is the fact that, according to modern physics, all processes cannot move faster than the speed of light. For a laptop of size $\approx$30 cm, this means that it takes at least 1 ns (10^{-9} s) for a signal to move from one side of the laptop to the other. During this time, even the cheapest laptops perform several operations. Thus, to speed up computations, we need to further decrease the size of the computer—and thus, further decrease the size of its memory units and processing units.

L. Bokati and V. Kreinovich, *Decision Making Under Uncertainty, with a Special Emphasis on Geosciences and Education*, Studies in Systems, Decision and Control 218,
https://doi.org/10.1007/978-3-031-26086-5_26

Need for quantum computing. Already the size of a memory cell in a computer is compatible with the size of a molecule. If we decrease the computer cells even more, they will consist of a few dozen molecules. Thus, to describe the behavior of such cells, we will need to take into account the physical laws that describe such micro-objects—i.e., the laws of quantum physics.

Quantum computing means reversible computing. For macro-objects, we can observe irreversible processes: e.g., if we drop a china cup on a hard floor, it will break into pieces, and no physical process can combine these pieces back into the original whole cup. However, on the micro-level, all the equations are reversible. This is true for Newton's equations that describe the non-quantum motion of particles and bodies, this is true for Schroedinger's equation that takes into account quantum effects that describes this notion; see, e.g., [2, 3].

Thus, in quantum computing, all elementary operations must be reversible.

Reversible computing beyond quantum. Reversible computing is also needed for different reasons. Even at the present level of micro-miniaturization, theoretically, we could place more memory cells and processing cells into the same small volume if, instead of the current 2-D stacking of these cells into a planar chip, we could stack them in 3-D.

For example, if we have a Terabyte of memory, i.e., 10^{12} cells in a 2-D arrangement, this means $10^6 \times 10^6$. If we could get a third dimension, we would be able to place $10^6 \times 10^6 \times 10^6 = 10^{18}$ cells in the same volume—million times more than now.

The reason why we cannot do it is that already modern computers emit a large amount of heat. Even with an intensive inside-computer cooling, a working laptop warms up so much that it is possible to be burned if you keep it in your lap. If instead of a single 2-D layer, we have several 2-D layers forming a 3-D structure, the amount of heat will increase so much that the computer will simply melt.

What causes this heat? One of the reasons may be design flaws. Some part of this heat may be decreased by an appropriate engineering design. However, there is also a fundamental reason for this heat: Second Law of Thermodynamics, according to which, every time we have an irreversible process, heat is radiated, in the amount $T \cdot S$, where S is the entropy—i.e., in this case, the number of bits in information loss; see, e.g., [2, 3]. Basic logic operations (that underlie all computations) are irreversible. For example, when $a \,\&\, b$ is false, it could be that both a and b were false, it could be that one of them was false. Thus, the usual "and"-operation $(a, b) \to a \,\&\, b$ is not reversible.

So, to decrease the amount of heat, a natural idea is to use only reversible operations.

How operations are made reversible now? At present, in quantum (and reversible) computing, a bit-valued function $y = f(x_1, \ldots, x_n)$ is transformed into the following reversible operation:

$$T_f : (x_1, \ldots, x_n, x_0) \to (x_1, \ldots, x_n, x_0 \oplus f(x_1, \ldots, x_n)),$$

where x_0 is an auxiliary bit-valued variable, and $\oplus$ denotes "exclusive or", i.e., addition modulo 2; see, e.g., [4].

It is easy to see that the above operation is indeed reversible: indeed, if we apply it twice, we get the same input back:

$$T_f(T_f(x_1, \ldots, x_n, x_0)) = T_f(x_1, \ldots, x_n, x_0 \oplus f(x_1, \ldots, x_n)) =$$

$$(x_1, \ldots, x_n, x_0 \oplus f(x_1, \ldots, x_n) \oplus f(x_1, \ldots, x_n)).$$

For addition modulo 2, $a \oplus a = 0$ for all a, so indeed

$$x_0 \oplus f(x_1, \ldots, x_n) \oplus f(x_1, \ldots, x_n) = x_0 \oplus (f(x_1, \ldots, x_n) \oplus f(x_1, \ldots, x_n)) = x_0$$

and thus,

$$T_f(T_f(x_1, \ldots, x_n, x_0)) = (x_1, \ldots, x_n, x_0).$$

Limitations of the current reversible representation of functions. The main limitation of the above representation is related to the fact that we rarely write algorithms "from scratch", we usually use existing algorithms as building blocks.

For example, when we write a program for performing operations involving sines and cosines (e.g., a program for Fourier Transform), we do not write a new code for sines and cosines from scratch, we use standard algorithms for computing these trigonometric functions—algorithms contained in the corresponding compiler. Similarly, if in the process of solving a complex system of nonlinear equations, we need to solve an auxiliary system of linear equations, we usually do not write our own code for this task—we use existing efficient linear-system packages. In mathematical terms, we form the desired function as a composition of several existing functions.

From this viewpoint, if we want to make a complex algorithm—that consists of several moduli—reversible, it is desirable to be able to transform the reversible versions of these moduli into a reversible version of the whole algorithm. In other words, it is desirable to generate a reversible version of each function so that composition of functions would be transformed into composition. Unfortunately, this is not the case with the existing scheme described above. Indeed, even in the simple case when we consider the composition $f(f(x_1))$ of the same function $f(x_1)$ of one variable, by applying the above transformation twice, we get—as we have shown—the same input x_1 back, and *not* the desired value $f(f(x_1))$.

Thus, if we use the currently used methodology to design a reversible version of a modularized algorithm, we cannot use the modular stricture, we have, in effect, to rewrite the algorithm from scratch. This is not a very efficient idea.

Resulting challenge, and what we do in this chapter. The above limitation shows that there is a need to come up with a different way of making a function reversible, a way that would transform composition into composition. This way, we will have a more efficient way of making computations reversible.

This is exactly what we do in this chapter.

26.2 Analysis of the Problem and the Resulting Recommendation

Simplest case: description. Let us start with the simplest case of numerical algorithms, when we have a single real-valued input x and a single real-valued output y. Let us denote the corresponding transformation by $f(x)$.

In general, this transformation is not reversible. So, to make it reversible, we need to consider an auxiliary input variable u—and, correspondingly, an auxiliary output variable v which depends, in general, on x and u: $v = v_f(x, u)$. The resulting transformation $(x, u) \to (f(x), v_f(x, u))$ should be reversible.

How to make sure that composition is transformed into composition. Let us fix some value of the auxiliary variable u that we will use, e.g., the value $u = 0$. We want to make sure that when $x = 0$, then in the resulting pair (y, v), the second value v is also 0, i.e., that $v_f(x, 0) = 0$. This way, $(x, 0)$ is transformed into $(x', 0) = (f(x), 0)$. So, if after this, we apply the reversible analogue of $g(x)$, we get $(g(x'), 0) = (g(f(x)), 0)$.

What does "reversible" mean here? In the computer, real numbers are represented with some accuracy ε. Because of this, there are finitely many possible computer representations of real numbers.

Reversibility means that inputs and outputs are in 1-1 correspondence, and thus, for each 2-D region r, its image after the transformation $(x, u) \to (y, v)$ should contain exactly as many pairs as the original region r.

Each pair (x, u) of computer-representable real numbers takes the area of ε^2 in the (x, u)-plane. In each region of this plane, the number of possible computer-representable numbers is therefore proportional of the area of this region. Thus, reversibility means that the transformation $(x, u) \to (f(x), v(x, u))$ should preserve the area.

From calculus, it is known that, in general, under a transformation

$$(x_1, \ldots, x_n) \to (f_1(x_1, \ldots, x_n), \ldots, f_n(x_1, \ldots, x_n)),$$

the n-dimensional volume is multiplied by the determinant of the matrix with elements $\dfrac{\partial f_i}{\partial x_j}$. Thus, reversibility means that this determinant should be equal to 1.

Let us go back to our simple case. For the transformation $(x, u) \to (f(x), v_f(x, u))$, the matrix of the partial derivatives has the form

$$\begin{pmatrix} f'(x) & 0 \\ \dfrac{\partial v_f}{\partial x} & \dfrac{\partial v_f}{\partial u} \end{pmatrix},$$

where, as usual, $f'(x)$ denoted the derivative. Thus, equating the determinant of this matrix to 1 leads to the following formula

$$f'(x) \cdot \frac{\partial v_f}{\partial u} = 1,$$

from which we conclude that

$$\frac{\partial v_f}{\partial u} = \frac{1}{f'(x)}.$$

Thus,

$$v_f(x, U) = v_f(x, 0) + \int_0^U \frac{\partial v_f}{\partial u}\, du = v_f(x, 0) + \int_0^U \frac{1}{f'(x)}\, du =$$

$$v_f(x, 0) + \frac{U}{f'(x)}.$$

We know that $v_f(x, 0) = 0$, thus we have

$$v_f(x, u) = \frac{u}{f'(x)},$$

and the transformation takes the form

$$(x, u) \to \left(f(x), \frac{u}{f'(x)} \right).$$

Examples.

- For $f(x) = \exp(x)$, we have $f'(x) = \exp(x)$ and thus, the reversible analogue is $(x, u) \to (\exp(x), u \cdot \exp(-x))$.
- For $f(x) = \ln(x)$, we have $f'(x) = 1/x$ and thus, the reversible analogue is $(x, u) \to (x, u \cdot x)$.

Comment. The above formula cannot be directly applied when $f'(x) = 0$. In this case, since anyway, we consider all the numbers modulo the "machine zero" ε—the smallest positive number representable in a computer—we can replace $f'(x)$ with the machine zero.

General case. Similarly, if we have a general transformation

$$(x_1, \ldots, x_n) \to f(x_1, \ldots, x_n) \stackrel{\text{def}}{=} (f_1(x_1, \ldots, x_n), \ldots, f_n(x_1, \ldots, x_n)),$$

we want to add an auxiliary variable u and consider a transformation

$$(x_1, \ldots, x_n, u) \to (f_1(x_1, \ldots, x_n), \ldots, f_n(x_1, \ldots, x_n), v_f(x_1, \ldots, x_n, u)).$$

To make sure that composition is preserved, we should take $v_f(x_1, \ldots, x_n, 0) = 0$. Thus, from the requirement that the volume is preserved, we conclude that

$$v_f(x_1, \ldots, x_n, u) = \frac{u}{\det \left\| \frac{\partial f_i}{\partial x_j} \right\|}.$$

Resulting recommendation. To make the transformation

$$(x_1, \ldots, x_n) \to (f_1(x_1, \ldots, x_n), \ldots, f_n(x_1, \ldots, x_n))$$

reversible, we should consider the the following mapping:

$$(x_1, \ldots, x_n, u) \to \left(f_1(x_1, \ldots, x_n), \ldots, f_n(x_1, \ldots, x_n), \frac{u}{\det \left\| \frac{\partial f_i}{\partial x_j} \right\|} \right).$$

26.3 Discussion

Need to consider floating-point numbers. In the previous text, we considered only fixed-point real numbers, for which the approximation accuracy ε—the upper bound on the difference between the actual number and its computer representation—is the same for all possible values x_i.

In some computations, however, we need to use floating-point numbers, in which instead of directly representing each number as a binary fraction, we, crudely speaking, represent its logarithm: e.g., in the decimal case, 1 000 000 000 is represented as 10^9, where 9 is the decimal logarithm of the original number. In this case, we represent all these logarithms with the same accuracy ε. In this case, the volume should be preserved for the transformation of logarithms $\ln(x_i)$ into logarithms $\ln(f_j)$, for which

$$\frac{\partial \ln(f_i)}{\partial \ln(x_j)} = \frac{x_j}{f_i} \cdot \frac{\partial f_i}{\partial x_j}.$$

In this case, formulas similar to the 1-D case imply that the resulting reversible version has the form

$$(x_1, \ldots, x_n, u) \to \left(f_1(x_1, \ldots, x_n), \ldots, f_n(x_1, \ldots, x_n), \frac{u}{\det \left\| \frac{x_j}{f_i} \cdot \frac{\partial f_i}{\partial x_j} \right\|} \right).$$

In some cases, the input is a fixed-point number while the output is a floating point number; this happens, e.g., for $f(x) = \exp(x)$ when the input x is sufficiently large. In this case, we need to consider the dependence of $\ln(f)$ of x.

Case of functions of two variables. If we are interested in a single function of two variables $f(x_1, x_2)$, then it makes sense not to add an extra input, only an extra output, i.e., to consider a mapping $(x_1, x_2) \to (f(x_1, x_2), g(x_1, x_2))$, for an appropriate function $g(x_1, x_2)$.

The condition that the volume is preserved under this transformation means that

$$\frac{\partial f}{\partial x_1} \cdot \frac{\partial g}{\partial x_2} - \frac{\partial f}{\partial x_2} \cdot \frac{\partial g}{\partial x_1} = 1.$$

For example, for $f(x_1, x_2) = x_1 + x_2$, we get the condition

$$\frac{\partial g}{\partial x_2} - \frac{\partial g}{\partial x_1} = 1.$$

This expression can be simplified if, instead of the original variables x_1 and x_2, we use new variables $u_1 = x_1 - x_2$ and $u_2 = x_1 + x_2$ for which $x_1 = \frac{u_1 + u_2}{2}$ and $x_2 = \frac{u_2 - u_1}{2}$. In terms of the new variables, the original function $g(x_1, x_2)$ has the form

$$G(u_1, u_2) = f\left(\frac{u_1 + u_2}{2}, \frac{u_2 - u_1}{2}\right).$$

For this new function,

$$\frac{\partial G}{\partial u_1} = \frac{1}{2} \cdot \frac{\partial g}{\partial x_1} - \frac{1}{2} \cdot \frac{\partial g}{\partial x_2} = -\frac{1}{2}.$$

Thus,

$$G(u_1, u_2) = -\frac{1}{2} \cdot u_1 + C(u_2)$$

for some function $C(u_2)$, i.e., substituting the expressions for u_i,

$$g(x_1, x_2) = \frac{x_2 - x_1}{2} + C(x_1 + x_2).$$

So, to make addition reversible, we may want to have subtraction—the operation inverse to addition; this makes intuitive sense.

Similarly, for $f(x_1, x_2) = x_1 \cdot x_2$, we get the condition

$$x_2 \cdot \frac{\partial g}{\partial x_2} - x_1 \cdot \frac{\partial g}{\partial x_1} = 1.$$

This expression can be simplified if we realize that $x_i \cdot \dfrac{\partial f}{\partial x_i} = \dfrac{\partial f}{\partial X_i}$, where we denoted $X_i \stackrel{\text{def}}{=} \ln(x_i)$. In these terms, we have

$$\frac{\partial g}{\partial X_2} - \frac{\partial g}{\partial X_1} = 1,$$

and thus, as in the sum example, we get

$$g(X_1, X_2) = \frac{X_2 - X_1}{2} + C(X_1 + X_2).$$

Thus, we get

$$g(x_1, x_2) = \frac{\ln(x_2) - \ln(x_1)}{2} + C(\ln(x_1) + \ln(x_2)),$$

i.e.,

$$f(x_1, x_2) = \frac{1}{2} \cdot \ln\left(\frac{x_2}{x_1}\right) + C(x_1 \cdot x_2).$$

So, to make multiplication reversible, we need to add a (function of) division—the operation inverse to multiplication. This also makes common sense.

References

1. O. Galindo, L. Bokati, V. Kreinovich, Towards a more efficient representation of functions in quantum and reversible computing, in *Proceedings of the Joint 11th Conference of the European Society for Fuzzy Logic and Technology EUSFLAT'2019 and International Quantum Systems Association (IQSA) Workshop on Quantum Structures*, Prague, Czech Republic, 9–13 Sept. 2019
2. R. Feynman, R. Leighton, M. Sands, *The Feynman Lectures on Physics* (Addison Wesley, Boston, MA, 2005)
3. K.S. Thorne, R.D. Blandford, *Modern Classical Physics: Optics, Fluids, Plasmas, Elasticity, Relativity, and Statistical Physics* (Princeton University Press, Princeton, NJ, 2017)
4. M. Nielsen, I. Chuang, *Quantum Computation and Quantum Information* (Cambridge University Press, Cambridge, 2000)

Appendix
What Is the Optimal Approximating Family

Several of our applications are based on common (or at least similar) mathematical results. These results are summarized in this special mathematical Appendix.

Comment. Results from this chapter first appeared in [1].

Need for approximations. In many practical situations, we need to find a good description of the observed data. In a computer, we can only store finitely many parameters. So, it is reasonable to consider finite-parametric approximation families, i.e., families that depend on finitely many parameters $C_1, \ldots, C_n$.

Need to consider families that linearly depend on the parameters. From the computational viewpoint, the easiest case is when the dependence on the parameters is linear, i.e., when the family has the form

$$f(x) = C_1 \cdot f_1(x) + \cdots + C_n \cdot f_n(x)$$

for some functions $f_1(x), \ldots, f_n(x)$. In this case, to find the values of the parameters C_i based on the known observations $x_1^{(k)}, \ldots, x_n^{(k)}, y^{(k)}$, it is sufficient to solve a system of linear equations

$$y^{(k)} = C_1 \cdot f_1(x^{(k)}) + \cdots + C_n \cdot f_n(x^{(k)}).$$

For solving systems of linear equations, there are efficient algorithms.

In principle, we can consider more complex dependencies—e.g., quadratic ones:

$$f(x) = \sum_i C_i \cdot f_i(x) + \sum_{i,j} C_i \cdot C_j \cdot f_{ij}(x).$$

However, in this case, to find the values of the corresponding parameters, we would need to solve systems of quadratic equations—and this is known to be NP-hard;

L. Bokati and V. Kreinovich, *Decision Making Under Uncertainty, with a Special Emphasis on Geosciences and Education*, Studies in Systems, Decision and Control 218,
https://doi.org/10.1007/978-3-031-26086-5

see, e.g., [2]. Thus, unless P = NP (which most computer scientists believe to be impossible), no general feasible algorithm is possible for solving such systems.

Since we want to have efficient algorithms, it is reasonable to restrict ourselves to approximating families in which the dependence on the parameters is linear.

Observations are usually smooth. Sensors usually smooth the observed processes, so what we observe is usually smooth. So, we can safely assume that the corresponding functions $f_i(x)$ are smooth (differentiable).

Thus, we arrive at the following definition.

Definition A.1 Let n be a positive integer. By an *approximating family*, we mean a family of functions

$$\{C_1 \cdot f_1(x) + \cdots + C_n \cdot f_n(x)\}_{C_1,\ldots,C_n}, \tag{A.1}$$

where the functions $f_1(x), \ldots, f_n(x)$ are fixed differentiable functions, and C_1, ..., C_n are arbitrary real numbers.

From this viewpoint, selecting a description means selecting n functions $f_1(x)$, ..., $f_n(x)$.

Towards the optimal description. Which description is the best? To answer this question, we need to be able to decide, for each two families of functions F and F', whether the first family is better (we will denote it by $F' < F$) or the second family is better ($F < F'$), or both families have the same quality (we will denote it by $F \sim F'$). Clearly, if F is worse than F' and F' is worse than F'', then F should be worse than F''. So, we arrive at the following definition.

Definition A.2 Let n be a positive number. By an *optimality criterion*, we mean the pair of relations $(<, \sim)$ on the set S of all possible n-dimensional approximating families that satisfies the following conditions:

- for every pair $F, F' \in S$, we have one and only one of the following options: either $F' < F$ or $F < F'$ or $F \sim F'$;
- for every F, F', and F'', if $F < F'$ and $F' < F''$, then $F < F''$;
- for every F, F', and F'', if $F < F'$ and $F' \sim F''$, then $F < F''$;
- for every F, F', and F'', if $F \sim F'$ and $F' < F''$, then $F < F''$;
- for every F, F', and F'', if $F \sim F'$ and $F' \sim F''$, then $F \sim F''$;
- for every F and F', if $F \sim F'$, then $F' \sim F$.

Definition A.2a. Let $(<, \sim)$ be an optimality criterion. We say that a family F is optimal with respect to this optimality criterion if for every other family F', we have either $F' < F$ or $F' \sim F$.

We want to use an appropriate optimality criterion to select a family. If a criterion selected several different families as equally good, then we can use this non-uniqueness to optimize something else. For example, if we have several different families that provide an equally good approximation, then, from all these optimal families, we can select, e.g., the family which is the easiest to compute. This additional selection is, in effect, equivalent to replacing the original optimality criterion with the new one $<_{\text{new}}$, according to which $F <_{\text{new}} F'$ if:

- either $F < F'$ according to the original criterion,
- or $F \sim F'$ and F' is easier to compute (in some formal sense, e.g., in terms of the computation time).

If the new criterion still selects several families as equally optimal, we can again modify it, etc.—until we end up with a *final* criterion for which there is exactly one optimal family.

Definition A.3 We say that an optimality criterion is *final* if it has exactly one optimal family.

As a starting point for measuring x, we can take different locations. If we select a different location which is x_0 units before the current one, then each new location x is identical to the old location $x' = x + x_0$. So, the same approximation that in the new units has the form $f(x)$ in the old units has the form $f(x + x_0)$. The relative quality of different profiles approximations should not change if we simply change the starting location. Thus, we arrive at the following definitions.

Definition A.4 For each family F as described by the formula (A.1) and for each x_0, by a *shift* $S_{x_0}(F)$, we mean a family

$$\{C_1 \cdot (S_{x_0} f_1)(x) + \cdots + C_n \cdot (S_{x_0} f_n)(x)\},$$

where $(S_{x_0} f_i)(x) \stackrel{\text{def}}{=} f_i(x + x_0)$.

Definition A.5 We say that an optimality criterion is *shift-invariant* if for every F, F', and x_0, the following two properties hold:

- if $F < F'$, then $S_{x_0}(F) < S_{x_0}(F')$;
- if $F \sim F'$, then $S_{x_0}(F) \sim S_{x_0}(F')$.

Similarly, nothing should change if we simply change the measuring unit for x—e.g., use miles instead of kilometers. If we replace the original measuring unit by a one which is λ times larger, then the new value x is identical to the old value $x' = \lambda \cdot x$. So, the same profile approximation that in the new units has the form $f(x)$ in the old units has the form $f(\lambda \cdot x)$. The relative quality of different profiles approximations should not change if we simply change the measuring unit. Thus, we arrive at the following definitions.

Definition A.6 For each family F as described by the formula (A.1) and for each $\lambda > 0$, by a *rescaling* $R_\lambda(F)$, we mean a family

$$\{C_1 \cdot (R_\lambda f_1)(x) + \cdots + C_n \cdot (R_\lambda f_n)(x)\},$$

where $(R_\lambda f_i)(x) \stackrel{\text{def}}{=} f_i(\lambda \cdot x)$.

Definition A.7 We say that an optimality criterion is *scale-invariant* if for every F, F', and $\lambda > 0$, the following two properties hold:

- if $F < F'$, then $R_\lambda(F) < R_\lambda(F')$;
- if $F \sim F'$, then $R_\lambda(F) \sim R_\lambda(F')$.

Proposition A.1 *For every n and for every final shift- and scale-invariant optimality criterion, the optimal family $F_{\rm opt}$ consists of polynomials of order $\le n-1$.*

Comment. This result is similar to results from [3].

Proof 1°. Let us first prove that the optimal family is shift- and scale-invariant, i.e., that $S_{x_0}(F_{\rm opt}) = R_\lambda(F_{\rm opt}) = F_{\rm opt}$ for all x_0 and λ.

Let us first prove shift-invariance of $F_{\rm opt}$. Since $F_{\rm opt}$ is optimal, for every family F, we have $F < F_{\rm opt}$ or $F \sim F_{\rm opt}$. In particular, this is true for the family $S_{-x_0}(F)$, i.e., either $S_{-x_0}(F) < F_{\rm opt}$ or $S_{-x_0}(F) \sim F_{\rm opt}$. Since the optimality criterion is shift-invariant, this implies that either $S_{x_0}(S_{-x_0}(F)) < S_{x_0}(F_{\rm opt})$ or $S_{x_0}(S_{-x_0}(F)) \sim S_{x_0}(F_{\rm opt})$. However, as one can easily check, we have $S_{x_0}(S_{-x_0}(F)) = F$. Thus, for every family F, we have either $F < S_{x_0}(F_{\rm opt})$ or $F \sim S_{x_0}(F_{\rm opt})$. By definition of optimality, this means that the family $S_{x_0}(F_{\rm opt})$ is also optimal.

Since the optimality criterion is final, there is only one optimal family, hence $S_{x_0}(F_{\rm opt}) = F_{\rm opt}$. Shift-invariance is proven.

Scale-invariance is proven similarly, by taking into account that for every F and every λ, either $R_{1/\lambda}(F) < F_{\rm opt}$ or $R_{1/\lambda}(F) \sim F_{\rm opt}$. So, by applying the scaling R_λ to both sides of these relations, we conclude that $R_\lambda(F_{\rm opt})$ is also optimal and thus, $R_\lambda(F_{\rm opt}) = F_{\rm opt}$.

2°. Shift-invariance means that every element of the family $S_{x_0}(F_{\rm opt})$ also belongs to the same family $F_{\rm opt}$. Let $f_i(x)$ denote the functions whose linear combinations (A.1) form the family $F_{\rm opt}$. Then, in particular, invariance means that for every i, the shifted function $f_i(x+x_0)$ is a linear combination of functions $f_j(x)$:

$$\begin{gathered} f_1(x+x_0) = C_{11}(x_0)\cdot f_1(x) + \dots + C_{1n}(x_0)\cdot f_n(x); \\ \dots \\ f_n(x+x_0) = C_{n1}(x_0)\cdot f_1(x) + \dots + C_{nn}(x_0)\cdot f_n(x), \end{gathered} \tag{A.2}$$

for some coefficients C_{ij} depending on x_0.

For each i, we can take n different values $x_1, \dots, x_n$ of x and get a system of n linear equations with n unknowns $C_{i1}(x_0), \dots, C_{in}(x_0)$:

$$\begin{gathered} f_i(x_1+x_0) = C_{i1}(x_0)\cdot f_1(x_1) + \dots + C_{in}(x_0)\cdot f_n(x_1); \\ \dots \\ f_i(x_n+x_0) = C_{i1}(x_0)\cdot f_1(x_n) + \dots + C_{in}(x_0)\cdot f_n(x_n). \end{gathered}$$

By Cramer's rule, the solutions $C_{ij}(x_0)$ to this system can be represented as a ratio of two polynomials in terms of $f_i(\cdot)$. Since the functions $f_i(x)$ are smooth, this implies that the functions $C_{ij}(x_0)$ are also differentiable functions of x_0.

Thus, we can differentiate both sides of (A.2) by x_0 and take $x_0 = 0$. As a result, we get a system of linear differential equations with constant coefficients:

$$f_1'(x) = c_{11} \cdot f_1(x) + \cdots + c_{1n} \cdot f_n(x);$$

$$\ldots \tag{A.3}$$

$$f_n'(x) = c_{n1} \cdot f_1(x) + \cdots + c_{nn} \cdot f_n(x),$$

where we denoted $c_{ij} \stackrel{\text{def}}{=} C_{ij}'(0)$.

The general solution to such a system is well-known (see, e.g., [3, 4]): it is a linear combination of terms of the type $\exp(\lambda_i \cdot x)$ and $x^k \cdot \exp(\lambda_i \cdot x)$, where λ_i are eigenvalues of the matrix (c_{ij}), and $k \le n - 1$ is a positive integer corresponding to the case when we have a multiple eigenvalue.

3°. Similarly, scale-invariance means that every element of the family $R_\lambda(F_{\text{opt}})$ also belongs to F_{opt}. In particular, this means that for every i, the re-scaled function $f_i(\lambda \cdot x)$ is a linear combination of functions $f_j(x)$:

$$f_1(\lambda \cdot x) = C_{11}(\lambda) \cdot f_1(x) + \cdots + C_{1n}(\lambda) \cdot f_n(x);$$

$$\ldots \tag{A.4}$$

$$f_n(\lambda \cdot x) = C_{n1}(\lambda) \cdot f_1(x) + \cdots + C_{nn}(\lambda) \cdot f_n(x),$$

for some coefficients C_{ij} depending on λ.

For each i, we can take n different values $x_1, \ldots, x_n$ of x and get a system of n linear equations with n unknowns $C_{i1}(\lambda), \ldots, C_{in}(\lambda)$:

$$f_i(\lambda \cdot x_1) = C_{i1}(\lambda) \cdot f_1(x_1) + \cdots + C_{in}(\lambda) \cdot f_n(x_1);$$

$$\ldots$$

$$f_i(\lambda \cdot x_n) = C_{i1}(\lambda) \cdot f_1(x_n) + \cdots + C_{in}(\lambda) \cdot f_n(x_n).$$

By Cramer's rule, the solutions $C_{ij}(\lambda)$ to this system can be represented as a ratio of two polynomials in terms of $f_i(\cdot)$. Since the functions $f_i(x)$ are smooth, this implies that the functions $C_{ij}(\lambda)$ are also differentiable functions of λ.

Thus, we can differentiate both sides of (A.4) by λ and take $\lambda = 1$. As a result, we get the following system of linear differential equations:

$$x \cdot f_1'(x) = c_{11} \cdot f_1(x) + \cdots + c_{1n} \cdot f_n(x);$$

$$\ldots \tag{A.5}$$

$$x \cdot f_n'(x) = c_{n1} \cdot f_1(x) + \cdots + c_{nn} \cdot f_n(x),$$

where we denoted $c_{ij} \stackrel{\text{def}}{=} C_{ij}'(1)$.

Here, for each i, we have

$$x \cdot f_i'(x) = x \cdot \frac{df_i}{dx} = \frac{df_i}{dx/x}.$$

Since $dx/x = d(\ln(x))$, we thus conclude that for the new variable $X = \ln(x)$ (for which $x = \exp(X)$) and for the corresponding functions $F_i(X) = f_i(\exp(X))$, we get the system of linear differential equations with constant coefficients:

$$F_1'(X) = c_{11} \cdot F_1(X) + \cdots + c_{1n} \cdot F_n(X);$$

$$\ldots \tag{A.6}$$

$$F_n'(x) = c_{n1} \cdot F_1(X) + \cdots + c_{nn} \cdot F_n(X).$$

Hence, similarly to the previous subsection, we conclude that each solution of this system is a linear combination of terms of the type $\exp(\lambda_i \cdot X)$ and

$$X^k \cdot \exp(\lambda_i \cdot X).$$

Substituting $X = \ln(x)$ into this formula, we conclude that each function $f_i(x) = F_i(\ln(x))$ is a linear combination of functions $\exp(\lambda_i \cdot \ln(x))$ and

$$\ln^k(x) \cdot \exp(\lambda_i \cdot \ln(x)).$$

Here, $\exp(\lambda_i \cdot \ln(x)) = (\exp(\ln(x))^{\lambda_i} = x^{\lambda_i}$.

Thus, each function $f_i(x)$ is a linear combination of functions x^{λ_i} and

$$\ln^k(x) \cdot x^{\lambda_i}.$$

4°. Our functions $f_i(x)$ are *both* shift-invariant and scale-invariant. Thus, each of them has to be both of form described at the end of Part 2 of this proof and of the form described at the end of Part 3. So, out of terms from Part 2, we cannot have exponential terms with non-zero λ_i—since these terms cannot be expressed in Part-3 form. Thus, the only possible terms are terms x^k with $k \le n - 1$.

So, each function $f_i(x)$ is a linear combination of such terms—and is, thus, a polynomial of order $\le n - 1$. The proposition is proven.

References

1. T.N. Nguyen, L. Bokati, A. Velasco, V. Kreinovich, Bhutan landscape anomaly: possible effect on himalayan economy (In View of Optimal Description of Elevation Profiles). Thai J. Math. 57–69. Special issue Structural Change Modeling and Optimization in Econometrics
2. P. Pardalos, *Complexity in Numerical Optimization* (World Scientific, Singapore, 1993)
3. H.T. Nguyen, V. Kreinovich, *Applications of Continuous Mathematics to Computer Science* (Kluwer, Dordrecht, 1997)
4. K.B. Howell, *Ordinary Differential Equations: An Introduction to the Fundamentals* (CRC Press, Boca Raton, Florida, 2016)

Index

L. Bokati and V. Kreinovich, *Decision Making Under Uncertainty, with a Special Emphasis on Geosciences and Education*, Studies in Systems, Decision and Control 218,
https://doi.org/10.1007/978-3-031-26086-5

V

W

Printed in the United States
by Baker & Taylor Publisher Services